ROUTLEDGE LIBRARY EDITIONS:
PHILOSOPHY OF RELIGION

A THREEFOLD CORD

A THREEFOLD CORD
Philosophy, Science, Religion

A DISCUSSION BETWEEN VISCOUNT SAMUEL
AND PROFESSOR HERBERT DINGLE

Volume 32

LONDON AND NEW YORK

First published in 1961

This edition first published in 2013
by Routledge
2 Park Square, Milton Park, Abingdon, Oxfordshire OX14 4RN

Simultaneously published in the USA and Canada
by Routledge
711 Third Avenue, New York, NY 10017, USA

First issued in paperback 2015

Routledge is an imprint of the Taylor and Francis Group, an informa business

© 1961 George Allen & Unwin

All rights reserved. No part of this book may be reprinted or reproduced or utilised in any form or by any electronic, mechanical, or other means, now known or hereafter invented, including photocopying and recording, or in any information storage or retrieval system, without permission in writing from the publishers.

Trademark notice: Product or corporate names may be trademarks or registered trademarks, and are used only for identification and explanation without intent to infringe.

British Library Cataloguing in Publication Data
A catalogue record for this book is available from the British Library

ISBN: 978-0-415-65969-7 (Set)
ISBN: 978-0-415-82935-9 (Volume 32)

Publisher's Note
The publisher has gone to great lengths to ensure the quality of this reprint but points out that some imperfections in the original copies may be apparent.

Disclaimer
The publisher has made every effort to trace copyright holders and would welcome correspondence from those they have been unable to trace.

ISBN 13: 978-1-138-96567-6 (pbk)
ISBN 13: 978-0-415-82935-9 (hbk)

A THREEFOLD CORD

PHILOSOPHY, SCIENCE, RELIGION

*A Discussion Between
Viscount Samuel
and
Professor Herbert Dingle*

London
GEORGE ALLEN & UNWIN LTD
RUSKIN HOUSE   MUSEUM STREET

FIRST PUBLISHED IN 1961

This book is copyright under the Berne Convention. Apart from any fair dealing for the purpose of private study, research, criticism or review, as permitted under the Copyright Act, 1956, no portion may be reproduced by any process without written permission. Enquiries should be addressed to the publisher.

© George Allen & Unwin Ltd 1961

PRINTED IN GREAT BRITAIN
in 11 point Juliana type
BY THE BLACKFRIARS PRESS LTD
LEICESTER

FOREWORD
by Lord Samuel

My partner in this venture wishes me to write the Foreword, and to explain the origin and purpose of this book and the reason for its unusual pattern as a dialogue. To do that I am obliged to begin with a brief piece of autobiography.

I was born in 1870, and until the age of fifty my interest was in politics and administration. Since then it has gradually shifted towards philosophy. That was caused by two events.

When the first world war broke out, in August 1914, I had been a member of the House of Commons for twelve years and of the Asquith Cabinet for five. No Government had ever been more devoted to the cause of international peace; but the treaty obligation resting on this country, guaranteeing the neutrality of Belgium and binding us to join in her defence against military aggression, was clear and inescapable. Government, Parliament and nation recognized that duty, regardless of the sacrifices it might bring. But that men should bring upon themselves so dreadful a calamity, in what was thought to be an age of enlightenment, showed that something must be fundamentally wrong with our civilization. It came as a moral shock to everyone, but to none more violently than to the twenty men who sat in the Cabinet room at Downing Street and had to bear the responsibility for the final decision.

The second event was personal. The war over, and the Turkish Empire in dissolution, the future of Palestine had to be decided. An international mandate conferred its administration on the United Kingdom, and I had the privilege of being appointed the first High Commissioner. I gladly accepted the honour of taking part in laying the foundation of a modern State in the ancient land of immortal fame. Twice in the history of mankind, on that soil had been engendered spiritual and cultural movements of supreme value to humanity. I held the belief—which I still hold —that Palestine in these latter days, might possibly add a third— on however modest a scale—to the religious revolutions of four thousand years and two thousand years ago.

But in the actual conditions, living for five years in Government House, then on the Mount of Olives over against the walled city of Jerusalem, I could see nothing to justify any vision of the kind. Jerusalem, the focus of three great Faiths, with its

Christian, Moslem and Jewish shrines, its Patriarchates and Bishoprics and Rabbinates, and its various pilgrimages, had by no means the atmosphere of a Holy City, radiating enlightenment, and peace on earth through the unification of mankind. On the contrary it was a seething centre of religious controversy and political conflict, between the Faiths themselves, between the several Churches into which the Faiths were divided, between the many sects, and, within the sects, between rival groups of dogmatic extremists. And I knew that the religious scene in Jerusalem was typical of what was happening in most of the countries of the world.

It was plain that the kind of politics to which we had all been devoted was not enough. It was the ideas underlying our policies and programmes that were at fault. Everywhere basic principles, religious and philosophic, were in a state of confusion. The impact of modern science on the ancient theologies had, for many, undermined religion itself. Philosophy was offering to the ordinary man no guidance.

Realizing this situation, and still governed by reaction from the war, I had to review my own beliefs and disbeliefs. I found them quite unsatisfactory, and decided that, when my term of office was over, I would withdraw from public life and think them out afresh.

This I did during the next ten years—in spite of some long interruptions from the political side. These brought me back in Parliament, and again, for a short time into Cabinet office; ending finally by relegation into the less exacting activities of the House of Lords. But during the ten years I had been able to make a wide survey, though necessarily superficial, of the situation as a whole; and to formulate such clear cut conclusions as I had been able to reach, in a short book, which I called *Belief and Action: an Everyday Philosophy*, published in 1937.

This was followed in 1951 by a tentative *Essay in Physics*, exploring the province where empirical philosophy overlaps with theoretical physics. It attracted the attention of Albert Einstein, with whom I had been honoured by a friendship of many years standing. I sent him a proof copy of my Essay, and he took the trouble to write me a long letter—concentrated on a single issue in which he was in disagreement—on the nature of Reality. This he allowed me to publish, in the original German and in translation, as an appendix to the book when it appeared. He said

FOREWORD

also that such a question could better be discussed in conversation than by correspondence, and suggested that if I was revisiting America I might come and see him in Princeton. This I was able to do, in 1952, and we had a long discussion, still limited to that one point, and ending with the disagreement not resolved, but rather emphasized by clarification.

I was unwilling to let so fundamental a matter rest there, and spent the next years in a further exploration, ending, in 1957, in yet another book, with the title *In Search of Reality*. This was well received by the reviewers but made no impact on the scientists, who it seemed did not feel called upon to pay attention to books about science written by non-scientists.

Such was the origin of this Dialogue, for it was the failure to get my conclusions across to the professional scientists that made me seek some different means of access. Among the few exceptions to the general indifference was Sir Henry Tizard, one of the most eminent of contemporary British scientists, whose status and the value of whose services in the second world war, are only now, after his death, becoming recognized.[1] He not only read the book, but induced several leading scientists to read it and to comment upon it. I consulted him as to the possibility of my undertaking a fresh presentation of the same main ideas in collaboration with a professional physicist of recognized authority, who would be in general sympathy with them but could subject them to expert criticism, and might reinforce those with which he agreed in language that the trained scientist—and philosopher—would understand and accept.

Tizard thought the idea a good one, and suggested as a collaborator Professor Herbert Dingle, lately retired from the Chair of History and Philosophy of Science at London University, and also from the Presidency of the Royal Astronomical Society. No name could have been more welcome to me, for Dingle had been my friendly guide for many years past, having given much time to reading the proofs and scripts of my books, lectures and broadcasts, and making valuable corrections and technical amendments, saving me from many mistakes. I put the proposal to Dingle, who at once accepted it, and set to work with enthusiasm. So this dialogue has come to be.

We devised for ourselves this procedure. First we agreed a plan

[1] As in the three articles about Tizard, from *Science and Government* by Sir Charles Snow, in *The Sunday Times* in March 1961.

for the book as a whole, dividing it into a number of sections. This plan was not rigid, it was changed as the course of the argument might require. Then, for each section, one of us would write an introductory sub-section, to which the other would contribute a continuation, bring out points of agreement, but formulating doubts, disagreements and alternatives. This usually led to a rejoinder, removing any misunderstandings that might have arisen, accepting amendments or reaffirming the original position. We might continue an interchange—still in typescript—on points of detail, before passing on to the next section. We had to resist many temptations to digress, and perhaps we have not always resisted those temptations as firmly as we should. The book has been written for the layman and the student, while not, by over-simplification, offending the expert and the erudite. But there may be some paragraphs here and there which the ordinary reader may find beyond him. If so, let him treat those pages, in a useful word of Asquith's, as 'skipworthy', and pass on—probably without losing the sense of the general argument. It will be found that, especially in the later sections, there are fundamental issues on which we have been obliged to register an irreducible disagreement. A little lively controversy may not be unwelcome, if it be no merely trivial argument over debating points. A society without controversies — political, economic, philosophic or religious—is a society intellectually moribund. But on the other hand, if none of the controversies are ever resolved and if no conclusions are ever reached—that is worse. It will soon be a dead society.

The reader will find that we are in fact in full agreement on many of the subjects that are discussed, and we hope that the conclusions may meet with approval from readers. On our main thesis—the urgent need for a greater association between philosophers and scientists, and of both with men of religion—there are happily signs at the present time of a marked trend in that direction. It is not suggested that those three great faculties of the human mind should merge their identities and abandon their specialized spheres of action; but only that they should evolve new ways of closer association—lending strength to the whole; like the strands of a rope, each of which is separate, keeping its own identity, but which, from beginning to end, are intertwined. As the Preacher says,[1] 'A threefold cord is not quickly broken'.

[1] *Ecclesiastes*, Chap. 4, v. 12.

CONTENTS

FOREWORD		page 7
I	View-Points	13
II	Energy and the Ether	51
III	Science and Mathematics	137
IV	Life and Mind	178
V	Retrospect and Prospect	214
APPENDIX I		265
APPENDIX II		270
INDEX		274

I

VIEW-POINTS

LORD SAMUEL:

1

Men's actions are governed by their ideas, but the ideas of our time, on all the matters of chief importance, are so confused that there is no body of belief, in any of the fields of thought and action, which is generally accepted. The theologies current in different parts of the world still offer, as they have always done, contradictory doctrines as the message of divine revelation; and now science has become predominant, and science teaches many things which conflict with dogmas of all the theologies. Meanwhile philosophy has gone its own way, with little regard either to the revolutionary conceptions of modern science or to obsolescent doctrines of ancient creeds. And the philosophers themselves, engaged in perennial controversies with one another, have been busy more with criticisms, refutations and negations than with any attempt to construct and establish a coherent body of rational thought.

The ordinary man seldom concerns himself with these high matters. He does not expect to understand them. Philosophy he usually finds repellent, theology dubious and remote. He waits for enlightenment and leadership from the intellectuals, whose business it is. But he waits in vain.

In consequence irrational, and often wicked, philosophies have sprung up. They gave birth to tyrannical political creeds, and to parties and systems of government violent and ruthless. So they have brought upon mankind great calamities; terrible wars and colossal crimes, worse than any recorded in all the bloodstained history of the human race. And still the lesson has not been learnt. Even worse is threatening. The nations of the East and of the West stand confronting one another in armed array, menacing the world with yet another war, rendered even more terrible by the nuclear bomb, with rocket-carriers making defence impossible.

Men's actions are governed by their ideas: I submit as the first proposition for our discussion that the time has come, and is indeed long overdue, when world opinion, through all its institutions and its agencies, should insist, as a first priority, that those three great faculties of the human mind—philosophy, science and religion—should no longer be content to remain at cross-purposes with one another and ineffective; but should combine their forces, and, each starting from its own standpoint, should join in building up that coherent body of rational thought, of which our civilization stands in such urgent need.

2

In that enterprise the initiative would seem to lie with philosophy. It occupies a middle position. On one side philosophy has contacts with natural science, particularly in the wide province of thought which empirical philosophy shares with theoretical physics. On the other side it has contacts with religion; for ethics has always been one of its chief concerns, while morality has held, and still holds, a foremost place in every religion. Further, if philosophy could succeed in establishing common ground with each of the others, it might also be able to help them in the end to overcome their mutual differences.

I hope you will agree so far. And I hope you will agree also that the philosophers have a duty, even more immediate, to try to clear up their own disagreements. For 2,500 years they have been engaged in earnest debates on the fundamental problems of the nature of the universe and of man's place in it. Many of the ablest minds that the human race has produced have devoted themselves to that perennial discussion. But, from all that effort, what result has followed?

No result has followed. The fundamental problems are problems still. No one can say—in even a single particular—'this is the answer that an accepted philosophy gives; this is the body of belief on which the world can build a cosmology—an organization of civilized communities—a code of ethics'—or whatever may be the subject of inquiry.

So barren has been the outcome of that strenuous effort so long continued that many, perhaps most, of the philosophers of the modern world have lost heart. They draw the conclusion that we shall never be able to draw any conclusion. Some try to escape into the byways of the meanings of words or the patterns of thought, and soon get lost in the mazes of linguistics or logic. Many students—especially among the most promising of the oncoming generation—are ruling out philosophy altogether, preferring to devote themselves to natural science as a more promising and rewarding study.

I do not think we need surrender to such defeatism, because a new line of approach to the old problems seems to be opening in a new environment. Some might dismiss as presumptuous anyone who can suppose that he has any new answers to offer to the riddles that have defeated all the greatest thinkers from the age of the Greeks to the nineteenth century, but the answer may be given that the world of the twentieth century itself is new. Ours is not the world that was known to Plato and Aristotle or Ptolemy, or to Spinoza or Descartes, or even to Kant and Hegel. Ours is the world revealed by the nineteenth-century astronomers and geologists, and by Darwin and Pasteur; the world of Marie and Pierre Curie, of J. J. Thomson and Rutherford; the world of radio and radar, of the great telescope at Mt. Palomar, and the new radio-telescopes at Jodrell Bank, Cheshire, and elsewhere; of our contemporary nuclear physicists, and of the young and eager sciences of biophysics and biochemistry.

Let us then, undeterred either by past failures or present diffidence—set out again along the old paths but now leading into new and different fields. And let us try to recapture the sense of mission and spirit of adventure that possessed the great pioneers, and with the hope that the far more powerful intellectual equipment of the present age may permit a greater success.

PROFESSOR DINGLE:

I am wholly in sympathy with the fundamentals of what you say, but I think there is a difference between us which may have important consequences, and we ought therefore to bring it into the open at the very beginning. I agree completely with your statement that philosophy, science and religion should not remain at cross-purposes with one another, and since, as an actual fact, each of them does now preserve a sort of independent existence, it would seem natural to try to reconcile them—which means, in effect. to create a comprehensive *Weltanschauung* (it is a pity that we have no satisfactory English word for this) in which each would be harmoniously related with the others. But immediately I contemplate this state of things it seems to wear such an air of unreality that I find myself forced to question the assumption on which it rests, and that leads me to the view that there are no entities corresponding to the names, philosophy, science and religion. Science, indeed, is so largely a matter of common agreement among scientists that on most points we can, without much error or indefiniteness, use such expressions as 'Science says . . . ' or 'That is unscientific', but even here there are some matters on which sharp divisions exist. But when we come to philosophy and religion, division is more obvious than unity. I cannot think it is possible that we shall ever attain to any expressible body of thought of which we can say in any universal sense, 'This is philosophy' or 'This is religion'. The differences of view on these matters, though many of them no doubt are due to misunderstandings or bad reasoning or prejudice or unconscious self-interest or what-not — all of which it is reasonable to hope are potentially removable—seem to me to reveal a residuum, and a large residuum, of essentially personal elements in the thought of each individual, which must always prevent universal agreement.

This seems to indicate a deep-lying difference in our tacit premises, and I think the difference can be expressed in this way —you will of course correct me if I misrepresent you. Behind your remarks there seems to lie the presupposition that there is some objective universe of things which is independent of us who can contemplate and study it in a purely objective way. By 'universe of things' I do not mean only the physical system which the astronomer calls the universe (though that is included, and may very well be taken as a fair sample, to which, for convenience, we might find it desirable, at least at first, to restrict our discussion), but some more comprehensive order, comprising moral laws, possibly as yet unapprehended spiritual existences, and so on. Our task is to discover the nature and character of this universe, and since it is quite independent of us it is conceivable that we might all agree about it. Indeed, if our investigations are free from error, we must do so, because it is what it is, no matter what we say or do or think. Science, philosophy and religion are distinguished merely by the fact that they concern themselves with different parts or aspects of this whole universe.

Their disagreements must denote errors on their part, for the universe itself cannot be two or more incompatible things at the same time.

I cannot admit this presupposition. I feel compelled to start building on what I know, what I cannot deny; not on what, however cogent the arguments for it may be, must remain an assumption. Indeed, one might argue from the fact that we have never reached agreement about this universe, that it is a chimera, but I do not wish to draw any conclusion of that kind. All that I feel compelled to acknowledge is that I must not start by assuming it; I admit the possibility that we might end by discovering it.

What I cannot deny is my own experience. I can deny any particular interpretation of it, but the experience itself remains. It is primarily a haphazard collection of all sorts of different things—particular sights, sounds, feelings, joys, sorrows, and so on—which come to me quite apart from my volition. My need for a *Weltanschauung* is the need I feel to make sense of this experience, to see it not as a chaos but as an ordered system in which every experience is rationally related to every other.

Of course, before I can advance from this primary starting-point to the point at which our present discussion must start—i.e. to the acknowledgment that you and I are fundamentally equivalent minds engaged in a common task of trying to elucidate commonly understood problems—I have to traverse a large amount of ground which it would be out of place to describe here. I have attempted it elsewhere,[1] and shall now assume the attempt successful. Although in the most intense experiences—those with which religion is concerned and the most intimate human relationships—I am forced back to the recognition that I must regard myself as unique, as I have no doubt you must do also, there is nothing that need arise here in which we may not speak of 'experience' as something common, or at least potentially common, to all human beings.

My substitute, then, for your problem of describing the nature of the universe which we experience, would be the problem of forming a description of a universe which makes sense of our experience. To take a very simple example of what I mean: where you would say, 'There is a star called Sirius, and it happens

[1] *Through Science to Philosophy* (O.U.P. 1937); *Solipsism and Related Matters* (Herbert Spencer Lecture, 1954; published in *Mind*, 64, 433, 1955).

that we can see it', I should say (if, of course, I were speaking in fundamental terms; I should not be so foolish as to use such language in ordinary conversation), 'I and others see a point of light in certain circumstances, and we express this fact by postulating something which we call a star and assign to a certain position in space'. One difference between these statements is that if (as is quite possible, though highly unlikely) astronomers should come to regard Sirius as a 'ghost star', I should simply have to make a slight change in my postulate whereas you would have to change the universe. But the fundamental reason why I prefer my statement is that it actually describes what we do, in the order in which we do it. We see the light first, and then proceed to the statement about the star: the star is a *conclusion* from the experience which is the *evidence* for it. Your expression implies not only that the star precedes the experience, but would have been 'there', and would have been exactly the same, if no one had ever experienced it.

In most scientific matters, of course, it comes to much the same thing whichever way we speak, and there is no point in exchanging familiar diction for pedantry: but even in science, I think, some of the points you raise in your books, *Essay in Physics*[1] and *In Search of Reality*,[1] go deep enough to make this difference of expression very relevant. But in philosophy, and especially in religion, it matters very much which way we speak. Here people do openly reason from experience to conclusion, and since each man's experience in these matters is largely peculiar to himself, the *Weltanschauung* which he reaches is also, in the last resort, peculiar to himself. That is why I do not think we can ever have a universally acceptable *Weltanschauung*—at least until we are all omniscient. We might, it is true, get nearer to one than we are at present, and this would certainly be worth doing, but at the best I think we are bound to fall far short of complete success.

It might be said, 'Very well; let us admit that a universal philosophy and religion are not likely to arise: we can still base science on the postulate of an objective universe'. But to admit that would be to sacrifice a unity which can be realized for the sake of approximating to one which cannot. I prefer to think

---

[1] Viscount Samuel—*Essay in Physics* (Blackwell, Oxford 1951, and Harcourt, Brace & Co., New York, 1952); *In Search of Reality* (Blackwell, Oxford, 1957, and Philosophical Library, New York, 1957).

that there is a possibility of universal agreement on what it is that we are doing when we try to form a Weltanschauung, although there is no possibility of a universal Weltanschauung, and since we cannot extend the 'objective independent world' idea to philosophy and religion, we must extend the 'rationalisation of experience' idea to science. That means that when we are concerned with the philosophy of science—as distinct from the actual prosecution of scientific research—we must discard the hypothetical objective 'universe of things' as a presupposition, and see what conclusions we can reach by acknowledging the fact that in *all* our thinking we accept the evidence of experience as our primary datum.

I believe, in fact, that science itself is becoming less and less amenable to description as the investigation of an external world by independent observers. I am not here referring to the much popularized idea that the distinction between objective and subjective in science has been eliminated, or at least blurred, by progress in quantum theory. That I regard as a gross error. What quantum theory does show is that the division between objective and subjective is not what we thought it was, but it is still as sharp as ever. If we insist on speaking of an objective external world and an observer who studies it, then the distinction has indeed become indefinite, but if we take what we experience (i.e. the observation, not the supposed thing observed) as the object presented to us, and our reasoning minds as the subjective element in the business, then the separation is still absolute. That is just the division which I have been advocating.

So I think that before we go any further we must clear up this question of an external universe as a primary postulate; otherwise we shall certainly be at cross-purposes.

LORD SAMUEL:

1

I am glad that you are able to begin with a point of agreement, and on a matter of importance—the need for seeking an understanding and association between philosophy, science and religion; instead of, as always hitherto and still at the present time, their continually emphasizing the points of divergence. But I confess to having felt a shock of surprise to read, immediately after, that you had found yourself driven to the view that 'there are no entities corresponding to the names, philosophy, science and religion'. If that is so, someone might ask how you were occupied during those years when you were Professor of History and Philosophy of Science at the University of London, and were indeed the first occupant of the first Chair established in this country with that name. He might ask also why you gave, to the comprehensive book which you published in the 1930's, the title *Through Science to Philosophy*—a track which I for one had already begun to follow, and have been encouraged to pursue ever since.

But that would be a misunderstanding. Clearly what you have in mind is a belief that there is not, and is never likely to be, a codex, so to speak, of ideas generally accepted by philosophers, by scientists and by men of religion. I remember reading—but I forget where — the shrewd observation, 'Behind every philosopher there lurks a man'. You hold that all three of these departments of human thought are pervaded by a personal element which will be constantly coming to the front to make agreement impossible. I allow that there is such an element, and also that we can never expect to reach finality; but I venture to hope that, piece by piece, generation after generation, we may go a long way in that direction. And I welcome your willingness to pool ideas with a desire to reach agreement on some at least of the root problems which have always vexed the minds of men, and which now, untreated and left to fester, infect our whole civilization.

So far so good. But the agreement has not lasted for long. You raise as a preliminary a point, which you rightly say, is of such basic importance that it must be cleared up before we can effectively get started.

2

Your point is in fact a re-opening of the oldest and most fundamental of the great issues in philosophy — the controversy between realists and idealists. I welcome your proposal to take this matter first—indeed I was going to suggest that we should give it priority among the questions to which we should address ourselves. And your statement of what you understand to be my position is quite correct (p. 16).

I do hold that we should begin by recognizing that there exists a universe, which is a given entity, real in its own right—the astronomical cosmos, with the solar system as an incident: included is the planet we name Earth, with all it carries on its surface—objects, processes, events; the human race, with their minds, their ideas, civilizations, institutions. This is one field, which men study and explore, being able to perceive some parts of it through their bodies, senses and minds, and through their reason, intuition, and, that strangest of all faculties, their imagination. All this is the universe as it is 'in itself'.

But this limited perception and portrayal has developed, in the course of human evolution, into a new, secondary, field of study—the universe as it is 'for us'. The process has of necessity been gradual. We cannot suppose that it would have been possible for simian man, emerging at the end of the Stone Ages as Homo Sapiens, to sit down within a generation or two and write a compendium of knowledge of the order of the Encyclopaedia Britannica. Knowledge had to begin with speculations, some of which developed through the ages into what we now call hypotheses, and some few of those into theories; later on, a small proportion became recognized as established laws of nature. The first speculations were mere guesses, as likely to be wrong as right; then and afterwards the hypotheses and theories had to be sifted into true and false. But what was meant by the expressions 'right', 'wrong', 'true', 'false'? What was the criterion? The answer was simple. The process was one of a long effort to bring 'the universe as it is for us' into correspondence with 'the universe as it is in itself'.

All this you would reject. Agreed that there were innumerable mistakes, you would attach chief importance to the mistakes: I would rather emphasize the rectifications. You say that those mistakes—and many more that no doubt still continue, unrecognized—make it impossible for us to put any confidence at all in those supposed representations of a real, given universe. You do not go so far as to say that any such thing is a chimera—although you think we might well do so; but you urge emphatically that we cannot accept it as fact.

3

With this thesis I am still in disagreement. And in the first place I cannot acquiesce in the recognition of a universe, real in its own right, being regarded as 'an assumption', or 'a presupposition', or 'a postulate', or 'an act of faith'. Those are question-begging words: if allowed to go unchallenged they would decide the argument in your favour before we had even begun to argue! Nor can I agree that the dichotomy, which we ought to make for purposes of a discussion such as this, should be defined as between 'an objective' universe and 'a subjective'; or even between a 'real' universe and an 'apparent'. Such definitions would confer an equal status on the two halves of the dichotomy, and seem to allow that a free choice is open between them. My contention is that there exists only one universe, 'which is what it is'. The other is the result of a valuable, and indeed essential, and largely successful effort to bring human ideas of nature into closer and closer correspondence with what nature in fact is. There we have our criterion. When we achieve a success we establish a new truth: when we fail we continue on a wrong road and sooner or later, in one way or another, we suffer accordingly.

And why should you think that your own experience is the one thing you are entitled to rely upon—enabling you 'to start building on what I know, what I cannot deny'? Why should that be held to be reliable, but the common sense of all mankind is not? And if you clear away everything except your own mind, how shall that escape? As Dean Inge said, 'The logical sceptic has no mind to doubt with'.

I have been reading lately Bertrand Russell's recent book *My Philosophical Development*[1] and I find there this passage:

'Under the influence of Idealist philosophers the importance of 'experience' has, it seems to me, been enormously exaggerated. It has even come to be thought that there can be nothing which is not experienced or experience. I cannot see that there is any ground whatever for this opinion, nor even for the view that we cannot know that there are things we do not know.'

Let me come now to the answers that I would submit to your fundamental objection to the realist position.

There are several answers, which supplement one another. First, no scientist would deny that, for immense stretches of time before man existed on this planet, the planet itself existed, together with the sun, and great numbers of other stars and of galaxies. To deny this would be to wipe out as worthless the

[1] George Allen & Unwin Ltd., 1959, p. 144.

whole body of fact discovered and established by successive generations of astronomers, geologists and anthropologists. But if it is not denied, then it must be admitted as true that for millions of years the human race did not exist but *a universe of some sort* did exist. When it is asked of what sort, we pass from the first field of inquiry to the second and we must consider that question in due course separately. Meanwhile we claim it to be established that there was a time when the cosmos existed, but the human race did not; there can be no question therefore of human perceptions, through sense-data or otherwise, then coming into the matter at all. In other words, during those epochs there existed a cosmos, real in its own right, and in no way dependent upon human characteristics, or accomplishments. Existence comes first; human perception comes after, if at all: and what is prior cannot be a construct of what is subsequent.

My second answer is that, without going back to distant epochs, we can find ample proof, all around us and every hour, that a universe, independent of human ideas, does exist—given and in its own right. Every living thing, animal or plant, is a proof. Each individual bird, beast, insect, fish, each individual tree, ear of corn, blade of grass, has come into existence, and can remain alive only through the existence of its environment. If no environment — no soil, no nutriment, no atmosphere, no earth, no sun—then that animal or that plant could not exist. But it does exist; therefore the environment must exist—and if an environment of some kind, then a universe of some kind, independent of human ideas—that is to say a universe real in its own right.

My third answer is independent even of animal and plant experience. Lifeless nature confirms the conclusion. We cannot doubt that the firmament of stars, this globe with its land, sea and atmosphere, the phenomena of light, heat and gravitation, are all facts in themselves, and in no way constructs of human minds.

4

But the idealist has another shot in his locker which, he may be confident, will dispose of all those arguments at a single discharge. Look, he will say, at the whole story of the development of human knowledge: it is one long record of empty myths and superstitions, misreadings of experience, mistakes and illusions.

Recall, at the very beginning, the animism of primitive man; the belief that the vagaries of climate and weather, the thunder and lightning, earthquakes, eclipses of the sun and moon, the motions of the stars and planets, were all the work of deities, demons, spirits of one kind or another. Only after thousands of years was it discovered, and after bitter controversies established, that all were governed by what we now call the laws of nature. The classic example of human error is of course the assumption, universally accepted throughout the ages until comparatively recently, that the earth is motionless and the centre of the cosmos, with the sun, moon and the whole firmament of stars and planets revolving around it every day. Recall again the belief, held as firmly and almost as long, that the material universe is built up of four elements—earth, air, fire and water; earth and water having an innate tendency to move 'downward', air and fire to move 'upward'. Or the belief that solid objects consist at bottom of vast numbers of very small indivisible atoms, which possess solidity, weight, colour and other qualities, as authenticated by the direct evidence of our own senses. Such is indeed the whole history of science—a continuous record of the detection and discarding of an unending series of false assumptions and wrong theories. Nor can it be supposed that this process happens to have been completed just now, in this second half of the twentieth century, and that there are no more errors still surviving. Is it not more probable that, fifty years hence, many of our present scientific beliefs will also have been rejected and thrown onto the vast heap of discards? And how can we tell in advance which of them it will be? It follows that we cannot have confidence in any one of them.

The conclusion, we shall be told by the idealist, must therefore be that it is a delusion to suppose that there exists any authentic and autonomous real universe, given and independent: or even, if it does exist, that human philosophers can concern themselves with it. The universe that we perceive, however imperfectly, by our senses, directly or indirectly, is the only one to which we can have access. We must make the best of it that we can; and any attempt to escape from that limitation by setting up a dichotomy such as has been proposed here, must fail for the reason that one of the halves is a mere figment.

The realist will agree that that historical picture is correct so far as it goes, but will not agree that it is the whole story. It brings in the mistakes and illusions but it leaves out the corrections and substitutions. For every old falsity that has been detected and discarded a new truth has been discovered and established.

And what do we mean when we speak, in this connection, of false and true, error and correction? Is there any referee, any judge? There is a referee, whose decision everyone accepts. Nature is the referee—for 'Nature is truth', as Meredith says.

5

I would give two illustrations, based upon facts which no one will dispute.

People usually carry a little diary in which to enter their engagements: in one of the preliminary pages of my diary for this year I find a list of 'Eclipses 1959'. From this—writing in the month of June—I learn that there is one more eclipse still to come in 1959, 'a total eclipse of the sun on October 2nd, visible as a partial eclipse in Great Britain'. Everyone accepts that prediction; but if someone were to doubt it, he would be referred to the known natural laws governing the movements of the earth and the moon in relation to one another and to the sun; he would be told of the centuries of experience in which there have been a great number of such predictions, all verified by the event. We are confident therefore that, when October 2nd comes, a shadow of the moon will sweep across the earth, and that, if we are at the right place at the right time and the sky is clear, anyone might see it doing so. Now contrast this with the beliefs about eclipses current among the Greeks and the Romans, and mentioned on many occasions by their historians. They held that they were portents sent by the gods, as a warning of some calamity about to happen or some danger to be avoided. Or the different belief held in China and in some parts of India—where, it is said, it still prevails—that a solar eclipse is the doing of some invisible demon or dragon, which is trying to destroy the sun, but which can be driven away by people gathering together and beating gongs and firing crackers.

My second illustration of the possibility of judging the truth or falsehood of the causes of natural phenomena, is taken from the history of medicine. In all tropical and subtropical countries malaria was an endemic disease; year by year it caused many deaths and debilitated vast numbers, depopulating whole provinces. It had been observed in Italy that people who lived in swampy districts, and particularly those who had to be out of doors at twilight or at night, were specially liable to this disease, and it had naturally been assumed that there was something noxious in the night air of those places; so the disease was given the name of 'malaria'. Such remained the situation, until Pasteur discovered that many human and animal diseases were caused by microscopic organisms which found entry into the bloodstream of individuals and destroyed its corpuscles. Next, an English doctor in the Indian Medical Service, Ronald Ross, hit upon the true solution. Swampy districts at twilight did indeed propagate the disease, but it was not because the air was contaminated: it was because the disease was bacterial and the infection was carried by a particular species of mosquito, and was transmitted by its bite. The insect flew at dark, and it could breed only by laying its eggs in stagnant water. Ross urged that if swamps and pools could be drained, or covered at the breeding season with a thin film of oil, which would be fatal to the mosquito larvae as they hatched out, the mosquitoes would be eliminated, and the malaria with them. So it proved. As we all know, the whole civilized world was alerted; millions of pounds or dollars were spent in campaigns against the anopheles mosquito, and now, wherever they have been successful, malaria has disappeared.

Anyone could quote a hundred similar cases of wrong theories replaced by better ones, from the history of preventive and curative medicine, and indeed from any one of the sciences.

'Vanity of vanities, saith the Preacher . . . all is vanity and vexation of spirit, and there is no new thing under the sun'—the achievements of human civilization are there to show how utterly false that is. And the well-known passage from Gibbon—'History is little more than the register of the crimes, follies and misfortunes of mankind': it is that indeed, but it must be very bad history if it does not register also greatness and virtue, achievement and well-doing, resplendent in the triumphs of reason and the victories of science.

6

This by the way. I return to my theme of the real existence of a given universe.

There remains one argument to the contrary not yet touched upon, which many idealists regard as the most convincing in their armoury and the most difficult for their opponents. While realists claim that modern science has come in to settle the ancient controversy in their favour, and particularly by its revelation of the inner structure of the atom, idealists have been asserting just the opposite. Now at last, they say, is it demonstrated that the 'naïve realism' of common sense is sheer illusion, for it is accepted as fact that the material universe is made up of atoms, each of which consists, not of a sphere of some kind of 'substance', but of a positively charged particle, or group of particles, with some neutrons without charge, at the centre, surrounded by one or more negatively charged particles or electrons all around it in some form of rapid movement—but the atoms being 'mostly volumes of emptiness'. It is therefore impossible to believe that it can at the same time be a collection of solid tangible objects, 'real in their own right', for the laws of nature do not permit two objects occupying the same space at the same time. The conclusion must be that it is a delusion to suppose that there is any real objective universe. There exists only a perceptual universe, which is the creation of our own minds. This was the contention of Sir Arthur Eddington; it was the principal lesson that he drew from the general acceptance of the Thomson-Rutherford theory of the structure of the atom. I would submit, with respect, that no such consequence follows.

I have been reading lately Sir Russell Brain's recent book *The Nature of Experience*, a reprint of three lectures delivered in May 1958. Combining the authority of an eminent practising neurologist with a philosophic outlook, he has given a helpful survey of these problems. In particular he confronts at the outset the point we are now discussing.

With his permission, I would quote Sir Russell Brain's statement of the case as presented by Eddington, and also by Bertrand Russell.

'Then there are Eddington's famous two tables.

' "One of them [he said] has been familiar to me from earliest years. It is a commonplace object of that environment which I call the world. How shall I describe it? It has extension; it is comparatively permanent; it is coloured; above all it is *substantial* . . . Table number two is my scientific table . . . It does not belong to the world previously mentioned—that world which spontaneously appears around me when I open my eyes, though how much of it is objective and how much subjective I do not here consider . . . My scientific table is mostly emptiness. Sparsely scattered in that emptiness are numerous electric charges rushing about with great speed; but their combined bulk amounts to less than a billionth of the bulk of the table itself . . . There is nothing *substantial* about my second table."

'Bertrand Russell, making broadly the same distinction as Eddington, draws the logical conclusion that Eddington's two tables must exist in two different spaces.

' "Naïve realism [he writes] identifies my percepts with physical things; it assumes that the sun of the astronomers is what I see. This involves identifying the spatial relations of my percepts with those of physical things. Many people retain this aspect of naïve realism though they have rejected all the rest. But this identification is indefensible. The spatial relations of physics hold between electrons, protons, neutrons, etc., which we do not perceive, and in the last analysis between coloured patches. There is a rough correlation between physical space and visual space, but it is very rough." '

Sir Russell Brain goes on to discuss the matter as a neurologist, considering especially the bearing upon it of experiences of illusions, hallucinations, apparitions and the like. He sums up his conclusions as follows (p. 24):

'The perceptual world, therefore, if I may use the term to describe the whole realm of our perceptual experience, is a construct of the percipient's brain.'

But he does not express any definite view on the problem illustrated by Eddington's two tables.

I think that realists may accept the situation as presented by Eddington, but need not agree to the deduction which he and others have drawn. A clue, that has the advantage of being quite simple and easily understandable, may lead to another conclusion.

7

If we accept, as we do, the reasoning that brings us to the two tables, we cannot avoid its carrying us much further. We may be surprised to find that it will reveal at least five tables—all of them physically real, all of them co-existing in the same place and at the same time. The explanation is that they exist at different size-levels.

Table No. 1 is at the size-level of the familiar world, of our own bodies and their environment. If we want any fresh evidence of the factual existence of our table as a real object in a real world, we might be able to call as a witness the carpenter who had made it, and who maybe might tell us how he had bought the wood from a timber-merchant who had imported a consignment from Sweden.

We pass on to an analysis of the material of which the table is made, but we shall not find ourselves at once confronted by Eddington's atoms. The wood consists of cellulose, and our Table No. 2 will be a structure of many millions of organic cells. A biologist who was a specialist in cytology could tell us a great deal about those cells: he would even be able to show us photographs of them, taken through a powerful microscope: he could describe the latest discoveries about their structure—their walls and the protoplasm which fills them, its highly complicated chemical composition, the nucleus which seems to direct their functioning. As to the real existence of those cells he would have no doubt at all.

The next size-level, where we shall find Table No. 3, is that of the molecules. These are far too small for the optical microscope, but the electron-microscope has now made it possible to take photographs of some of the largest molecules that are found in organic material—the proteins. These are highly complicated structures, one molecule sometimes consisting of thousands of atoms, kept together by binding electric forces. We can find in the text-books pictures of imagined models of protein molecules.

It is only at the next size-level that we come to the atoms which gave Eddington his Table No. 2 but give us our Table No. 4. These are far beyond the possibility of vision, direct or indirect; but we have enough evidence of their reality in the great chemical industries, which give employment and livelihood to millions of workpeople in all the more advanced countries of the world, for atoms are their raw material. Combinations of a few or many of the chemical elements—those discovered now number nearly a hundred—are the foundation of those industries. And if it is suggested that atoms do not exist as real objects in a real world we would ask how then could molecules exist; and if no molecules, then there could be no cells, no cellulose, no wood, no table.

Far below the size-level of the atom is that of the particle. Our Table No. 4 was an organized arrangement of atoms; each atom has been discovered to be a system of various kinds of particles; this gives us our Table No. 5. The factual reality of the sub-atomic particles has been more often challenged than that of the other entities that we have been discussing. It was at first suggested that particles might prove to be nothing more than mathematical quantities in the differential equations of physicists. It was even thought by some that it would be proper to describe them as 'waves of probability'. But it is only necessary to visit one of the great new atomic research establishments which are handling such particles to find conclusive evidence that they are real 'things'. For at Harwell, for example, one can see the elaborate precautions that have had to be taken, at vast expense, to protect the staff from the danger of being exposed to streams of neutrons or electrons, or even from single, or a few, free radio-active particles. One can see there the concrete walls several feet thick and faced with heavy slabs of lead, that have been erected around the parts of the plant where such particles are being dealt with; for, as all the world now knows, such contamination is exceedingly dangerous—may cause the disease of leukemia, or cancer of the blood cells, or other kinds of serious, or even fatal, injury to the human body. Only a few days ago I saw in an illustrated paper a photograph of the widow and orphan son of one of the Japanese fishermen, whose boat, in a remote part of the Pacific, had been struck by a fall-out of radio-active particles from a test of nuclear bombs by the American navy. That Government had recognized its liability for the injuries done to the crew, and had paid such compensation as money can give to the victims or their families.

My point is that all these risks and injuries, precautions and compensations and the like, are proof positive of the physical existence of sub-atomic particles. A differential equation cannot give us cancer: 'waves of probability' cannot kill fishermen.

8

I said that if we are persuaded by Eddington and others to accept a theory of two tables we should find ourselves obliged, by the same reasoning, to accept *at least* five. At a size-level immensely far below that of the particles, Max Planck discovered what he named 'the quantum of action', usually expressed as 'Planck's constant h'. This has revealed another world at the level, in relation to our standards of measurement, of 10^{-27}. The new science of quantum physics has come into being: but it has not yet established any conclusion as to the status of the quantum. Is it purely mathematical? Or is it 'something'—an entity, discrete and indivisible, leading to a theory that the universe, at bottom, is granular? In that case we should have a sixth table. The present stage is still one of study and discussion among physicists.

But there is one other group of subjects which we have not yet touched upon, and which we may postpone but cannot omit: it is another of the old problems—the problem of the nature and processes of life and mind. This is becoming again a live issue in the light of the new psychology and of the researches of the biochemists and biophysicists. It is indeed the subject of active contemporary discussion and controversy, under the heading of the physical basis of mind, or the mind-body relation. But we have agreed that we cannot discuss everything at once, and I think that it would be best to postpone this to a later stage of our dialogue.

PROFESSOR DINGLE:

1

Let me first clarify one or two subsidiary points before coming to the main issue. I have no objection to the terms 'idealist' and 'realist' if they are used merely as convenient labels for our respective views as we express them here, but these words have been given so many different shades of meaning in philosophical discussions that I would not in general accept 'idealist' as descriptive of my position. This proviso is necessary because philosophers have been known to place an opponent in a recognized class, and then to 'refute' him by arguments against that class which in fact have no application to his ideas. I am sure you would not do so, but a tacit acceptance of the title 'idealist' would expose me to a risk which I do not care to run.

Secondly, I do not attach greater importance to the 'mistakes' than to the 'rectifications'. I regard any hypothesis or theory later than the first as, at the same time, both a rectification and a mistake—a rectification of an earlier theory and a mistake which a later theory will rectify.

Thirdly, I do not place greater reliance on my experience than on 'the common sense of all mankind'—far from it. But, in the fundamental sense in which I speak of 'my experience', the common sense of all mankind becomes an inference from it, for I know of, say, Newton's law of gravitation only because of my experience of reading about it, hearing it spoken of, and so on, and I have to form from those sights and sounds an idea of what the law is. As I said, the problem of justifying the use of the term 'experience' as descriptive of 'the common sense of all mankind' is an inescapable one, but one which, for reasons of space, I must here regard as solved; but in what I shall have to say in a moment about the double meaning of such terms as 'existence' and 'time', there will be at least a clue to the double meaning of 'experience'.

2

Coming now to the main subject, your distinction between the universe as it is 'in itself' and as it is 'for us' opens up the possibility of a common starting point, for I am perfectly ready to admit a universe as a valid concept, though at bottom a hypothetical one. It may be, of course, that there is no 'universe', our experiences falling into two or more radically distinct classes: as a particular case, it may be that the material and the mental sciences are each potentially perfectible in themselves, with no regular relation between them. But I think it is the faith (though not, as is sometimes said, a necessary assumption) of most scientists at least that it is possible to bring the whole of experience into a single system, and that system, though we have as yet only the most rudimentary glimpse of what it might be, could be called a 'universe'.

By contrast there are the various 'world pictures' which are put forward from time to time by scientists and philosophers, which represent what, on the basis of our present knowledge, they conceive this universe to be. Ignoring for the moment the differences between contemporary thinkers, and assuming for simplicity that the science of any one epoch permits only a single such 'world picture', it is a fact of history that the world picture does change from epoch to epoch as knowledge grows. We may therefore designate the current world picture as 'the universe as it is for us'.

So I would admit your distinction if 'the universe as it is in itself' stands for what, if we were omniscient and had unlimited and infallible reasoning power, we should describe as 'the universe', and 'the universe as it is for us' stands for what, with the knowledge and reasoning power we have at any assigned time, we at that time describe as 'the universe'. Such a distinction certainly facilitates expression and involves no intellectual arrogance so long as we remember that 'the universe as it is in itself' is an article of faith, and that we must never allow our investigations to be perverted or restricted by the assumption, conscious or unconscious, that its postulation is a necessary condition of our thinking.

But I am afraid, from the arguments which you give for the recognition of 'the universe as it is in itself', that that is what you would maintain. Let me call 'the universe as it is in itself' the 'real' universe, and 'the universe as it is for us' the 'apparent' universe—merely, of course, for brevity and without opening the door to any metaphysical problems concerning 'reality'. Then the three answers to the 'idealist' in your paragraph 3 (p. 23) are arguments not merely for the recognition of a real universe —which, with the meaning given above, I should be prepared to grant—but for its *independence of experience*. To you the real universe exists quite apart from any actual or possible observation of it, and would exist—indeed, at one time did exist —in exactly the same sense as it does now, if there were no one to observe it. To me the real universe must be defined ultimately in terms of its relation to experience; otherwise I see no justification for defining it at all. (I would remark parenthetically that your arguments are strictly applicable only to what we usually call the physical universe. I do not know whether you would hold that, for instance, moral laws as well as the law of gravitation existed before there was any life, and I do not want to complicate the discussion by intruding that question if the general problem can be dealt with equally well in terms of the physical world alone; but, in view of possible future developments, I think we should be clear about the limitations of our present considerations. That being understood, we may for the present keep to the actual facts which you cite).

Of your three answers I think the first is the most crucial one, and the only one that poses a really fundamental problem. The second argues that if animals and plants exist, then non-living things exist in the same sense, which I readily grant: the third is more an assertion of the conclusion than an argument for it. But the first demands very close examination. Its essential point is that, according to generally accepted scientific opinion (which, to prevent misunderstanding, I at once admit that in this respect I wholly share), the material world is older than any form of life. The conclusion is then that the existence of the material world is independent of the existence of life or mind, and therefore must be accepted as an inescapable fact to which all our thinking must conform, and not merely as a permissible hypothesis facilitating the correlation of experience.

My reply to this is that the word *existence*—and by implication the idea of *time*—is used here in two different senses. I can perhaps best explain my meaning by an analogy—and it is merely an analogy, not an example of that which it is employed to elucidate.

In *The Tempest* Prospero undoubtedly exists, but the actors whom he calls into play in Act IV are mere phantoms; they do not exist. This distinction is quite clear and, granted the tacit premises from which I start and which I am sure you automatically accept, you will agree with me. But now, if I change those premises and compare Prospero, not with the actors but with Shakespeare, then I am sure we shall both agree that Shakespeare exists and Prospero does not. So, depending on our implied premises, we can say with equal truth that Prospero exists and Prospero does not exist.

It is the same with experience and the material universe. In the sense in which experience exists the material universe does not, but within the material universe itself it would be perfectly legitimate to say that stars exist and gremlins do not. The material universe is postulated as a means of expressing the inter-relations of experience. As experience grows our description of it changes, so that the view of the universe which we hold today is vastly different from that of the Middle Ages, and no one can say what changes will be made in the future.

Nor is this merely a change of description of something whose *existence* remains unquestioned. A single example will suffice to show this. You will remember Edmund Gosse's fascinating account of his father's attempt to reconcile the findings of geology with the Scriptures by the postulate that the world was created at about the same time as man, with the fossils inserted ready-made in the rocks. If that postulate is accepted, the argument from the long ages of a lifeless universe breaks down. I am not, of course, pleading for Gosse's hypothesis. I only point out that it is a *permissible* one, and that means that there is a possibility, however faint, that it may be true. How could this be discovered? Only by the discovery of facts not at present known i.e. by the acquisition of fresh experience—which would make this the most acceptable hypothesis to hold.

But this means that the hypothesis of a pre-human universe is at the mercy of experience, and that is fatal to the argument that it is independent of experience. If there were, in your sense of the word, an existent universe before any intelligence had begun to appear, it would be impossible for that intelligence, when it did appear, to disprove that fact. Yet, unless the whole basis of scientific investigation is discarded, it is not so impossible. Hence such an existent universe cannot be admitted. There is no escape by discarding the basis of scientific investigation, for in that case there would be no ground even for the *hypothesis* of a pre-human universe.

Of course I do not question that—within the implied but rarely expressed 'universe of discourse' — we must grant the existence of stars at times on the physical time-scale earlier than the first appearance of life on the Earth, but that time-scale is itself a postulate which we can and do alter at will for the rationalisation of experience. It is not something given us with which we cannot tamper. What we call time when we are meditating on our most elementary experience of continued existence is something very different from the scientist's time-scale in which events are placed in an order and the intervals between them are allotted measure-numbers. The former is indeed inescapable, but the latter is a purely mental creation subject to unlimited change as experience grows. The difference is brought home to us most vividly, I think, when we ask ourselves what we mean by time going backwards. What is usually understood by this is the reversal of the order of events—the sort of thing we should see if a cinematograph reel were put through the machine from the wrong end. But a reversal of our unavoidable sense of time would be a state in which we had no memory of the past, but remembered the future until it came upon us, when it would immediately vanish from our knowledge except in so far as we could recover it by calculation just as we now calculate future eclipses. The difference between these two conceptual happenings indicates the profound difference between time in the 'given' sense of the word and the time-scale of science.

Now your argument for a real universe is based on the fact that there are events which are placed before the advent of life and consciousness in the time order of the latter, but it does not take account of the fact that that time order is itself a creation of our rational minds for the sole purpose of bringing order into our experiences, and as experience accumulates it is subject to an unlimited amount of change, and even to complete destruction, if our demand for rationality requires it.

This point is so fundamental that I venture to present it in yet another aspect. Suppose we ask the question: who created Frankenstein's monster? There are two possible answers: (a) Frankenstein; (b) Mary Shelley. Both are correct within their respective contexts, but we must not mix them. We must not credit Mary Shelley with practical engineering ability or hold her responsible for the monster's misdeeds; nor, on the other hand, can we ascribe Frankenstein's genius to inheritance from Mary Wollstonecraft.

We are, in fact, in no danger of doing this because we understand quite clearly that Mary Shelley stands wholly outside the novel, and that conditions which pertain to the novel alone are not conditions to which she is in any way subject. We do not ordinarily recognize this distinction so clearly when for the novel we substitute the scientific world-picture (the apparent universe) and for Mary Shelley our reasoning minds, but in fact we have created the scientific world-picture just as truly as Mary Shelley created the novel. There is this difference, of course, that what controlled the details of her creation was her sense of what the public would read, or her feeling for art, or the desire to teach a particular lesson, or whatever her dominant motive may have been, whereas what controls the details of the scientific picture is the scientist's need to order his experiences into a rational system. But in each case the creation is wholly at the mercy of the creator, who is subject only to his chosen purpose and the material provided him (in the scientist's case the latter is, of course, that which we are given in experience).

There will doubtless be a large part of our discussion to which all this will be irrelevant, in which we shall be entirely within the apparent universe, but it is necessary, I think, for our respective view-points to be delineated as clearly as possible at the outset. From what I have said, my attitude to the contents of your paragraphs 4 to 8 will be fairly obvious, but I think it will be helpful if I comment on some of the specific points you have raised.

3

The theme of paragraphs 4 and 5 is the fact of what we call scientific progress. When we exchange one idea for another we proceed from worse to better, and history shows that this process has been continuous. With this, of course, I entirely agree. What I must dissent from is the statement that this progress is an advance from error to truth in the ordinarily understood meaning of the terms: 'For every old falsity that has been detected and discarded a new truth has been discovered and established'. You have chosen examples which indeed make this appear plausible, but I do not think it is possible to describe the history of science in that way.

There is only one thing that makes uninterrupted progress in science, and that is the extent to which experience has been found to be inter-related. The theories we hold, the world-pictures we form, the concepts we employ—all these things indeed change, but show no *ordered advance*; they sway to and fro, they are dropped and picked up again, what was 'false' becomes 'true' and then 'false' again, and no one knows what now outmoded idea might not become the accepted 'truth' of tomorrow.

We once thought that elements could be transmuted; we then 'discovered' that that was an illusion, the indestructible and unchangeable molecules of each primary substance remaining eternally 'unbroken and unworn' in Maxwell's phrase; today we are back at the idea of transmutation, and claim that we have actually achieved it; tomorrow . . . ? We once thought space was filled with an 'ether'; we then found there must be several ethers; then we realised that there was none at all; today we are rather undecided, but many physicists think it possible that there is 'a sort of ether'; tomorrow . . . ? We once thought light was a feeler put out by the eye; we then saw that it must be something that enters the eye from without; we found it necessary to describe this something as a collection of particles; next, this was 'disproved', and light was indubitably a wave-motion; today some try to hold both views together and others claim that you can hold either provided that you do *not* hold both together; tomorrow . . . ? We once thought the Sun moved in a circle; then we found that it was at rest and the Earth moved; then we saw that the Sun did move after all, but in a straight line; next, it did really move in an orbit, but a very large one; today we think that it is all a matter of convention anyway, and we can take our choice; tomorrow . . . ?

So we could go on. The point about this 'progress' is not merely that the ideas change, but that they *drift*; there is no system in the change. Not only is it wrong to denote each change as a change from falsity to truth; it is not even an advance towards the 'truth'. We cannot extrapolate from the direction of change to the truth ahead because there is no direction of change.

But what is achieved at every step is the discovery of a new relation between experiences. The change from corpuscular to undulatory light was accompanied by the establishment of a relation between the velocity and the refraction of light which was unknown before and was not conveniently expressible in the old terms. The partial reinstatement of the corpuscular idea is accompanied by the establishment of a relation between light and certain electrical phenomena. And so on. And this is inevitable because it is the establishment of such relations between experiences that is the whole object of science, and unless a change of ideas augmented such relations it would not occur. These scientific concepts are not objective truths; they are instruments that facilitate the discovery and expression of relations between experiences, and they are taken up and modified and discarded solely in order to further that end.

Let me revert to Philip Gosse. If science is the discovery of the truth behind things, then he is not only unanswerable but he has given the complete and final answer to the scientific problem. By one single hypothesis—that of a God who wills certain things and has the power to achieve them—everything is explained. Whatever happens, it is because he wills and does it. It is an eminently legitimate hypothesis. We know of beings with wills and executive powers; we merely postulate one such being who possesses these faculties in exceptionally high degree, and that is all. Everything that happens is explained by one single hypothesis of an indubitably valid kind. William of Ockham would have been enchanted.

Why, then, did nobody accept the hypothesis? Simply because, whatever they may have imagined they were doing, what the scientists were really seeking was not the *explanation of* the things that they observed, but *rational relations between* the things that they observed. God willed that the apple should fall and that the Moon should encircle the Earth, but science was not asking why these things happen; it wanted to know how the fall of the apple was related to the orbit of the Moon, and it was the discovery of that relation that constituted scientific progress. We have had various theories of why gravity occurs, but nobody is interested in them so long as they do not bring more experiences into the scheme of relations. Whether they are 'true' or 'false' is not a matter of any scientific importance.

4

The essence of your paragraphs 6 to 8 is, I think, this. We can give quite different accounts of the 'objective universe', exemplified by Eddington's account of his two tables. You depict an idealist who concludes from this that there is no 'objective universe', that which is so called being a multi-valued creation of our own minds. You admit the possibility of giving different accounts of the universe—indeed, you add others to Eddington's —but you do not admit that this justifies the conclusion that there is no objective universe. If I understand you rightly, you regard your table 5, which is composed of sub-atomic particles, as being the 'real' table, for you regard the existence of sub-atomic particles as proved by such facts as the possible occurrence of cancer in those working in atomic energy establishments. 'A differential equation,' you say, 'cannot give us cancer.'

But here again I think we must distinguish the facts given us in experience from the ideas which we form in order to relate them together. The fact is that the experiences involved in having cancer are found to follow exposure to certain pieces of scientific apparatus. We at present express this relation by postulating that particles produced in the apparatus enter our bodies and bring about certain changes there. That expression is justified, of course, because it fits in with a large number of other facts, and so is not a mere *ad hoc* explanation but a part of a much wider scheme of correlation of experience.

It is not the only possible interpretation of the facts. We could think of many others, just as you have given several interpretations of your experiences with the table. The one which we choose is not the 'true' one; we do not know which, if any, that is. We choose, in fact, the simplest one that is sufficient for our immediate purpose. Thus, in ordinary everyday affairs we choose Eddington's first table; in certain departments of theoretical physics we choose his second. In the case of cancer the variety of choice does not appear so obvious because we have not the same variety of attitudes to the phenomenon. We have usually only one aim with regard to cancer—to prevent or to cure it—and we therefore express it in the terms most likely to serve that end.

But that does not constitute 'proof positive of the physical existence of sub-atomic particles', even within the purely physical world (the world of the novel), let alone in the world of mental existence (the world of the author). The time may very well come when the hypothesis of atoms and sub-atomic particles will be found unequal to the task of correlating all that we shall know, and some continuous conception of matter may take its place in which 'sub-atomic particles' will have no more meaning than 'celestial spheres' have in modern celestial dynamics. If you accept the spheres, then you can regard the irregularity of eclipses as proof positive that they are not coaxial, but if you don't, the same fact has a quite different implication in which the spheres have no meaning. In just the same way, if you accept the existence of sub-atomic particles, then their undesirable effects on the human body can be regarded as proved—it is the necessary way in which the facts must be interpreted in terms of those entities. But that does not in the least prove that you must accept their existence.

The statement that a differential equation cannot give us cancer is of course (with a reservation mentioned below) true, but no one would suppose that it could. All that is meant when it is said that you can represent a particle only by a differential equation (I am neither accepting nor denying that assertion) is that no verbal statement about something picturable to the eye can be a true description of the particle; its behaviour in assigned circumstances can be determined only by solving such an equation and interpreting the result. That does not mean that the equation is the particle.

But in a more fundamental sense, I think your epigram does overlook a vitally important point. It is not to be taken for granted that a differential equation cannot give us cancer. We must set no limits whatever to the possibilities of experience—apart, of course, from those demanded by logical necessity. In the past men have certainly believed that the pronouncement of magical formulae could induce disease, and it may be that this was by no means an illusion. Granted an uneducated savage believing implicitly in the powers of a witch doctor, it is not inconceivable that a differential equation, uttered by the latter after a solemn warning that it was carcinogenic, might indeed be so. I am no psychologist, and I make no assertion on the point, but it is decidedly not a possibility which could be rejected *a priori* if facts pointed in that direction.

What I mean is that, if we are to be true scientists, all conceivable facts of experience, however improbable, must be allowed to be possible. By the same token, all schemes of rationalization (e.g. all possible descriptions of a table) must be regarded as *prima facie* permissible, and rejected absolutely only if they involve contradiction, and temporarily only if they are not so convenient as some other for the particular purpose in view. We cannot prove the unique truth of any one of them by an appeal to experience; all that can be proved in that way is that it is or is not valid as a possible conception within a limited field of experience.

But, to repeat, these basic considerations are not likely to be pertinent to a large part of our discussion. They are basic, and it is therefore necessary that they should appear at the beginning, but we may reasonably hope that we shall usually be on a higher level, from which an appeal to them need be rare.

LORD SAMUEL:

1

I must answer two or three specific points that you have raised before I revert yet again to your main proposition.

I have concurred with the postponement of the problems of Life and Mind, and suggest that we should devote a later part of our dialogue to that group of subjects. I think that questions of dreams, incantations, literary fictions and the like should be taken up in that context, and feel no doubt that you will concur.

As to those tables—I have not said, and do not hold that the fifth or any one of my five should be singled out as 'the real table'. Each one in turn is the only one existing at its own size-level, and is real as such. If we could grow smaller and smaller like Alice in Wonderland, and plunge down deep after deep to the successive levels, at once we should find ourselves in a world consisting of molecules; then of atoms; and then of particles. Of course these are not to be imagined as separable from one another, like a 'nest of Chinese boxes', perhaps eight or ten of them, round, of diminishing sizes, made to fit perfectly, one inside the other, that can be separated and set out side by side. But if one of my imagined tables were to disappear, if for example, the wood and its cells were burnt, the molecules of which they were made would no longer exist as a structure with the overall shape of table number one; with the molecules would be dispersed their atoms, and the particles together with the atoms.

But you are right in questioning (p. 40) my assertion that the protective measures at Harwell and the tragic fate of the Japanese fishermen are 'proof positive of the physical existence of sub-atomic particles'. I still think that it is right as far as it goes: I confess, however, that I had not in mind the wave-particle controversy, which has been active among physicists during the last thirty years, and has still not reached a satisfactory conclusion: and, further, that the shortest waves in the Maxwell electromagnetic waveband, the gamma-rays, may produce some of the same phenomena as radio-active particles; among them the harmful effects upon the human body. But I think my assertion would hold good if it were worded that the safety precautions necessary at atomic research establishments, and events such as the disaster to the Japanese fishermen, are 'proof positive' of the real existence of some *physical factor* able to cause such effects.

I have only one other point in your second contribution to which I would wish to offer a reply before commenting at greater length on your incidental reference to our familiar conception of time. With respect to my illustration about the eclipses, I cannot be content that you should be unwilling to use any stronger word than 'plausible' when comparing the present accepted explanation of a solar or lunar eclipse with the explanations given by the Greeks and Romans or by the Chinese. It seems to me a grave matter when a former President of the Royal Astronomical Society is reluctant to speak of 'true' and 'false' in such a connection. And in general, the whole conception of human experience as the impregnable foundation of all our knowledge seems to me wrong. The man-centred cosmology of the pre-Copernican astronomy is now unanimously recognized to have been a mistake. Are you sure that the man-centred universe of the idealists is not a repetition in philosophy of exactly the same mistake?

A few years ago The Times was guilty of a minor *gaffe* which caused some amusement at the time. In its number of June 30, 1954, in the usual list of 'Arrangements of the Day', there had been included an entry 'Total Eclipse of the Sun'. The incongruity of ranking a majestic event in the heavens with the parties, banquets, conferences and other events of the London season was a piece of human arrogance which could only be excused as a sub-editor's slip. Or would you say that philosophically it could be justified? From your observation just now about Sirius (p. 17), one might think that you would. After all, we were not responsible for the eclipse and could do nothing about it except look at it. What mattered was not the event—which in fact did not happen where we imagined we saw it, but where the Sun had been eight minutes before. The readers of The Times needed to be told that an eclipse would take place on that day, in order that, if they wanted to look at it, they should remember to fit it in with their engagements. Quite right, therefore, to include it among the 'Arrangements of the Day'!

Don Marquis satirized the man-centred universe not unfairly[1]

> the supercilious silliness
> of this poor wingless bird
> is cosmically comical
> and stellarly absurd.

2

Your reference to time was in the course of your reply to my thesis that there exists a universe, real in its own right and in no way dependent upon human perceptions or ideas; and that

---

[1] *Don Marquis, archys life of mehitabel*, p. 96 (Faber & Faber, 1934).

science and philosophy must study both that cosmos and also the man-centred world which is the outcome of those perceptions and ideas. Wherever the second is found to differ from its counterpart in the first, it is the business of science to try to bring them into conformity with one another. This can be done—has in fact continually been done, and is being done, piece by piece, more successfully today than at any previous time—along either of two lines of approach: either by making new discoveries about the cosmos and its processes; or else by further research within our own field of knowledge, resulting in fresh adjustments in our own ideas about nature: more often it is done along both lines simultaneously. To this you answer that we have no proof that the first exists, but, if it does, our minds can have no access to it except by way of the second.

If I have understood rightly what you say in the passage I am referring to (p. 33)—and it is very possible that I have not—our dialogue seems now to be taking this form:

In support of my thesis I cite the long ages when man did not yet exist but the earth and the solar system, and much else, did exist; and I say that the cosmos cannot then have been consequential, or in any way dependent upon human perceptions and ideas, because what is earlier cannot be consequential or dependent upon what has come into existence.

You say (p. 34) that you 'wholly share the generally accepted scientific opinion that the material world is older than any form of life'. And you go on to draw a conclusion that the concept of the existence of the material world is independent of that of the existence of life and mind.

I do not agree that that follows: and in any case it seems to me that you are raising there a new and separate question, which will certainly have to be considered, but only when we come to discuss the general subject of life and mind. But keeping to my limited and simple point, I would claim at this stage, that our premiss, which is agreed by us both—that the material universe existed before human perceptions existed—at once refutes the basic principle of the idealist philosophy. This is that the man-centred world is the only basis for a man-created philosophy, and that a cosmos, 'real in its own right', is either non-existent or inaccessible.

It is at this point that you bring in the question of time.

You say (p. 35), 'Of course I do not question that—within the implied but rarely expressed "universe of discourse"—we must grant the existence of stars at times on the physical time-scale earlier than the first appearance of life on the Earth, but that time-scale is itself a postulate which we can and do alter at will for the rationalisation of experience. It is not something given us with which we cannot tamper. What we call time when we are meditating on our most elementary experience of continued existence is something very different from the scientist's time-scale in which events are placed in an order and the intervals between them are allotted measure-numbers. The former is indeed inescapable, but the latter is a purely mental creation subject to unlimited change as experience grows.'

That argument implies that we can only consider events or processes that are arranged on a time-scale. But time-scales are, as you say, the invention of human philosophers and scientists: in those primordial ages there were no philosophers and scientists, and therefore there were no time-scales. Consequently you and I are unable to take into account anything that was happening then. It follows that my argument is inadmissible *a priori*, and the idealists are entitled to ignore it, and to stand by their own hypothesis as though it had never been challenged.

But is not this a good example of the confusions into which we find ourselves plunged if we refuse to make that clear differentiation, for which I have been pleading, between 'the universe as it is in itself' and 'the universe as it is for us', but are constantly fluctuating between the two? For here at one moment we are envisaging the cosmos when it existed but men did not—as you agree was once the case; but the next moment we intrude upon it with 'time-scales', which we cannot leave out, because you say they are 'inescapable'. but which we ought not to bring in, because they were not then existing.

From my standpoint 'time-scales' are nothing more than a philosophic contraption. We can discard them as easily as we have created them. They are part of the great structure of mathematics—a marvellous achievement of the human intellect, but belonging wholly to the man-centred world. The cosmos does not give us measurements, or numbers, or any other of the conventionally agreed symbols. The cosmos gives us nothing more than the fact that events do not happen all at once but follow one after another—one was then, another is now—and we may infer that others will happen later. So also with what we term 'space': events—or objects—are not situated together—but one is here, another is there, others are elsewhere. When we wish to measure the intervals between any two of them—intervals in sequence or intervals of extension—we have to depend upon figments of our own making: 'spaces'—on measurements that may be based perhaps upon our own bodies, the length of the thumb, the arm, the foot, a thousand paces: 'times'—upon the processes of nature, the revolutions of the earth on its axis, of the moon round the earth, of the earth round the sun. But 'space' in itself, and 'time' in itself, are fictional abstractions.

3

This will sound to most people like an empty paradox. It never occurs to us to doubt the reality of space and time. We measure spaces with yardsticks and milestones, we measure times with sundials or watches or calendars. All this is indeed 'inescapable': we could not possibly dispense with the words 'space' and 'time' from our vocabularies. Why then should we bring into question the reality of a generalized space and a generalized time of which these particularized spaces and times are parts?

Poets, orators and preachers have personified time and exalted the vastness of the universe: they seem to take a pathetic satisfaction in contrasting the insignificance of man with their infinity and eternity.

> Time, like an ever-rolling stream,
> Bears all its sons away.

Or the great lines of Andrew Marvell:

> But at my back I always hear
> Time's winged chariot hurrying near.
> And yonder all before us lie
> Deserts of vast eternity.

The treasures of the poetry of nature we shall never cease to cherish. Nor is there any question of the growing influence of philosophy and science requiring us to do that: it is the privilege of the arts that they may transcend fact and soar on the wings of imagination. But, on the other hand, the arts must not demand of philosophy and science that they should subordinate reason to imagination, fact to fancy. As Dr Johnson wrote, 'We may take Fancy for a companion, but must follow Reason as our guide'.

4

Modern physics rightly lays chief stress on observation and experiment; it is uncompromising in forbidding unverifiable inferences and unsupported assumptions. That has been made very plain in the long drawn-out controversies of the present century on the principle of uncertainty and on the wave-particle relation. But when we come to this question of the reality of space and time, do physicists observe their own rule, or even take it into account? The student will often find in text-books of physics, at the very beginning of the discussion of fundamentals, some such phrase as 'The universe having been created in a space-time framework'; with an argument following to establish —what is now generally agreed—that if space and time are accepted they cannot be considered as separate entities but only as integrated units of space-time; that there are in nature no points and instants, but only point-instants. That is all well and good: but it does not touch the question whether there is any space, whether there is any time: for if there is not, there cannot be any space-time either, and all that part of Einstein's theory of relativity which is built upon the acceptance of a space-time continuum as the substratum of the cosmos and the seat of all phenomena, is invalidated.

How, then, does the insistence upon observation and experiment, as the only road to the ascertainment of truth, apply here? What research scientist has ever observed space as such, or has experimented in his laboratory with time? Here is yet another example of the necessity to distinguish between the cosmos as it is 'in itself' and the man-centred world as it is 'for us'; for when we try to go on from the particular spaces and times, to which we are accustomed, to discover a correspondence with a generalized abstract space and time and space-time, which we can bring in as the ultimate basis of the physical cosmos, we find that no such correspondence exists. This brings us at once to the basic question of the status of mathematics in philosophy. For space, time and space-time are parts of the great mathematical structure which is one of the masterpieces of human achievement—of reason and imagination in close alliance.

PROFESSOR DINGLE:

1

I think our views on fundamentals are now clearly set out, and I do not wish to hold up the progress of our discussion with points of relative detail, important though they may be. I would therefore merely make the brief remark that it is only when we are at rock bottom, so to speak, that I would hesitate to say that the present explanation of eclipses is 'true'. In any other connection, such hesitation would of course be merely captious, but the problems we are concerned with are fundamental, and it is because I think it likely that particular questions will arise in which we shall be forced down to rock bottom that I want at the outset to establish myself there.

Whether an eclipse is an 'arrangement' is, I suppose, largely a verbal matter. But suppose that, with the progress of knowledge, we find ourselves impelled to postulate that the solar system is an organism, in which a sort of super-mind bears the same relation to the movements of the planets as our minds to the circulation of our blood. We have now come to regard many bodily processes, which appear entirely physical, as mental in origin, and we describe and, if they are diseases, treat them accordingly. It is therefore possible that a time will come when what we shall call the 'true' explanation of eclipses will be a psychological one, and the present account will become as trivial as a geometrical description of the path of the blood through the arteries and veins.

A somewhat similar idea was, in fact, held a hundred years ago by Fechner—not a mere visionary, but a reputable man of science—but I am not now concerned with its probability; its possibility is all that I would urge as a justification for denying the attribute of ultimate truth to the present description of eclipses. And indeed I should go further. Even if no such alternative were imaginable, I should still hold it wrong to say of any 'explanation' that we might give of any fact of experience, that it is the final truth about that experience. Whatever account we give, I do not see that we can ever know that a better is not possible.

2

I think we are more in agreement on time-scales than you seem to imply. From my standpoint also, ' "time-scales" are nothing more than a philosophic contraption,' though I would hardly agree that 'we can discard them as easily as we have created them'. In the present state of knowledge I do not see how we could even begin to tackle philosophical problems without speaking of events that are arranged on some sort of time-scale. But that is comparatively superficial. Fundamentally we are at liberty to modify, or even abandon altogether, the whole conception of time-scales if further experience should make that seem the proper course.

What I said was inescapable was not 'the scientist's time-scale' but 'what we call time when we are meditating on our most elementary experience of continued existence' — that which makes it impossible for me to avoid recognizing a distinction of some sort between the vision of my room which I have now, and the vision, 'in the mind's eye', of the Grand Canyon in Arizona, say, which also I have now, whereas in ordinary parlance I last saw the Grand Canyon in 1933. We at present make that distinction by calling the latter vision a 'memory' of an 'event' which we postulate and locate at an earlier point on our invented time-scale than the event (also postulated) of seeing the room. This is a very convenient way of acknowledging the two facts that my *experiences* of seeing the room and the Canyon are simultaneous, and that they are essentially of different character. I place both *experiences* at the point marked 'now' on my time-scale, and the *events* which I postulate as causing them at different points—the point now and the point 1933, respectively—on this same time-scale.

While we are discussing physics at least, we shall not need to bother about experiences, for, although they are the origin of the whole business, physics does not begin until we have assigned them to events which we regard as objective. Thereafter we do not talk of my reading of the thermometer or of your reading of the thermometer but of 'the reading' of the thermometer. So I think we may agree—at least until we come to discuss questions of morals—that time and space are entirely man-made concepts which must be our slaves and not our masters in our attack on physical problems. I think this modifies somewhat the summary of our positions which you give on pp. 44 ff.

I would, however, add one footnote to these remarks, in case our discussion leads us into regions in which I may have to appear to contradict them. A bad habit has developed almost unconsciously in modern physicists, which leads them, in effect, to regard their datum not as 'the reading of the thermometer' but as the reading which a hypothetical observer would make if he were subject to certain aberrations which they impose on him. I have known cases—and they are not exceptional but almost normal—in which, when purely objective problems of a certain type are presented to physicists, they instinctively turn a measuring-rod or a clock into an 'observer', and discuss what he would do in the circumstances of the problem if he were constrained to act like certain symbols in an equation.

This might be legitimate as a short cut if it were possible to reach the same result by a longer but purely objective route, but, in the problems in question, that is not possible; the result claimed can be reached only by animating the apparatus. It would be legitimate also if the problem were discussed in terms of the *actual* observations made—e.g. in terms of my reading of the thermometer instead of 'the reading', provided that everything that might distinguish my reading from a reading on which all physicists could agree as correct, were determined and allowed for. But that would be a clumsy and altogether unnecessary process which no sensible person would think of adopting so long as we confine ourselves to purely physical problems into which psychology does not enter.

When, therefore, I say that in discussing physics we need discuss only the objective events, I must admit that many physicists at the present day (it is quite a modern phenomenon) do not do so. If occasion arises this malady can be exemplified, but since neither of us is likely to become its victim, it is not worth while to hold up the discussion on this point.

II

ENERGY AND THE ETHER

LORD SAMUEL:

1

During my lifetime there have been developments in science, both theoretical and applied, which have revolutionized physics. I remember vividly the excitement throughout the world when Marie and Pierre Curie, following upon the pioneer work of Becquerel, succeeded in isolating a radio-active element, to which they gave the name of radium. Already Hertz had extended our knowledge of the electromagnetic wave-band by the discovery, beyond the infra-red, of the long waves that became the basis for sound and visual broadcasting, and for radar; and Röntgen had discovered, at the other end beyond the ultra-violet, what are termed the X-rays. The discovery of the very short and penetrating gamma-rays came as a concomitant of radium. So the door was opened to J. J. Thomson and Rutherford's discovery of the electron and of the inner structure of the atom itself.

No less revolutionary were the experiments, in 1887 and the following years, of the American scientists Michelson and Morley, undertaken with the object of checking the measurements by the astronomers of the velocity of the Earth's movement through the ether. These had given the surprising result that the velocity was nil, in other words that no such movement existed. But since it had been established, beyond challenge, that the Earth does in fact move round the Sun, physicists gradually came to the conclusion that the null result of Michelson and Morley proved that no such thing as an ether existed. It did indeed show that a quasi-gaseous ether, such as Newton had imagined, and which had already been brought into question on quite different grounds, did not exist. But it did no more than that. Nevertheless physicists showed their readiness to welcome new ideas, however destructive they might be, by discarding some of the fundamental theories of Newtonian mechanics altogether, and even abandoning all attempts to frame a realist cosmogony; turning to mathematics and its symbolism to furnish a substitute. Einstein's special and general theories of relativity were offered to fill the vacuum, and they have dominated the situation ever since. The discovery by Max Planck of a new fundamental unit of energy—the quantum of action—came as a supplement: but the status of the quantum has not yet been determined.

Some time later came the discovery of what is termed the red-shift in the spectra of distant stars and galaxies, and this has been accepted by most physicists, perhaps a little hastily, as involving a cosmic theory of an expanding universe—stars and galaxies rushing apart in every direction with colossal velocity.

2

It is necessary to recapitulate—briefly as it must be—the doubts and objections that have arisen on several important points in present-day physics, and to suggest, if possible, alternative lines of approach. Starting from the standpoint of the co-operation between philosophy and science—to which you assent, and for which you have indeed long been in this country one of the principal protagonists—I should be greatly interested to learn what measure of agreement, if any, you can give to these suggestions.

I will take Michelson-Morley first, for that was the real origin of the present confused situation.

I well remember a brief, but significant, observation on this matter by Einstein in conversation long ago. It was in 1922, in Jerusalem, where I was holding the post of High Commissioner for Palestine under the British Mandate, and Einstein was on his way back to Europe after a lecture-tour in Japan, and was staying with me. He had come to deliver the first lecture of the embryo Hebrew University of Jerusalem, taking as his subject 'The Meaning of Relativity'. With the friendly tolerance that he always showed to any well-meaning layman anxious to learn about a subject that was unavoidably abstruse, he was willing to talk about relativity on our walks in the grounds of Government House and along the rocky paths of the surrounding slopes of the Mount of Olives. He described how relativity had been the consequence of the failure of Michelson and Morley to trace any movement of the Earth through an ether, and the remark that I remember was 'If Michelson-Morley is wrong, then relativity is wrong'.

I do not think that anyone, scientist or philosopher, would have contended at that time that Michelson and Morley were wrong, in the sense that their ingenious and elaborate apparatus was badly designed for its purpose, or that their mathematics was faulty; or, indeed that the negative answer to the question that they had put was mistaken. It is the same now. The experiments have been repeated again and again, in every kind of environment, and although one researcher has believed that he had been able to detect some slight movement, scientists in general, now as throughout the last half-century, agree that Michelson-Morley is not wrong—so far as it goes. But I ventured to submit, in the previous book that has been mentioned, and would still maintain—that the whole undertaking was based upon a tacit assumption, and that that assumption was mistaken.

It was assumed—as a matter of course and not established by argument—that, if there was an ether, the undoubted movement of the Earth was *in* and *through* it. That implied that the Earth was one thing and the ether was another; the Earth was situated *in* and was moving *through* the ether, as fishes are in and move through the sea, or birds the air. The large and elaborate apparatus of Michelson and Morley—with its rods, mirrors, prism and interferometer—being physically attached to the surface of the Earth, was assumed to be one factor in the situation; and the flashes of light that were emitted and reflected, being not so attached, belonged to the other factor, the ether. It followed that the one could be located relatively to the other, and the relation measured mathematically.

But this notion is incompatible with the essential character of an ether, namely that it is a universal continuum—a plenum, which exists, if at all, everywhere and always; leaving no interstices or empty spaces, resembling not at all the water of the sea, with the fishes in it, or the gases of the air with the birds. It would therefore be as true to say that the ether is *in* the Earth as that the Earth is *in* the ether: consequently there cannot be any movement of the Earth through the ether because there is no place that it could move into in which ether was not already present. It might have been foreseen from the beginning that any experiment of the kind designed by Michelson and Morley was bound to fail. What was wrong was the initial assumption—hidden, ignored, but tacitly implied in those innocent looking prepositions, 'in' and 'through'.

If this is accepted, it will be seen that the conclusion to be drawn from that negative result is not that no such thing as an ether can exist, for it is quite possible to imagine a continuum, of a different kind altogether, to which that objection would not apply.

Indeed, some eminent men of science have never acquiesced in the view that the whole question had been disposed of by Michelson-Morley, and in recent years their number has received important additions. I must fortify my case by quoting the gist of what they have said: at the same time I do not want to delay the course of the argument by burdening it with some pages of extracts: I have therefore relegated them to an appendix (Appendix 1).

It will be seen that I have been able to cite Sir Arthur Eddington, Sir Oliver Lodge, Sir J. J. Thomson, Albert Einstein, Professor P. A. M. Dirac, Lord Cherwell and Sir Edmund Whittaker. The mere recital of those names should be enough to make a discussion of ether once more respectable. An eminent contemporary physicist, who was good enough, at the invitation of the late Sir Henry Tizard, to read my recent book and to send me some valuable notes upon it, says on this point 'On the general question of an aether I think most physicists would agree that opinion is tending towards it; but,' he adds, 'it will have to be a special sort, very unlike the kind Oliver Lodge used to be so fond of'. With that reservation we must agree. And I was glad to note that you yourself say in the previous section (p. 37), that 'today we are rather undecided, but many physicists think it possible that there is "a sort of ether" '.

3

But anyone is entitled to say: if you assert that an ether forms part of the given universe, and if you suggest that it may be something quite different from the quasi-gaseous medium which Newton supposed it to be but which was knocked out by Michelson-Morley, what kind of thing do you suggest that it is and how does it work? That question is entirely legitimate. As

Oliver Lodge wrote, 'The ether is a physical thing ... Its mechanism is unknown to us; yet a mechanism it must have, for it is subject to physical laws.'

Those who turn to my supplementary note (Appendix 1) will observe that many of the writers who are quoted bring in energy as the main factor, and that may well be, if not the answer, at least part of an answer to the question. For wherever in physics we probe phenomena to the furthest point we can reach, we find ourselves ultimately face to face with energy—and with nothing else. Whether it is waves of electromagnetic radiation, or electrons or other sub-atomic particles, or the motion of material objects, or now the quantum (which is a quantum of *action*), or electricity, or gravity—push our analysis as far as we can and it is energy, and that alone, that we have left at the end. Is it not possible that the point at which man's analysis stops is the point where nature's synthesis begins?

But to use the word energy, and to say, with J. J. Thomson, that 'the seat of all phenomena' is there, does not give us any hint of a theory as to the common origin of all those varied phenomena, or their mechanism. It would not give us the ordered universe that we know.

The task we have set ourselves is very difficult because of the limitations that we have accepted. It is not to be any sort of gas 'more subtle than air'. But also it must not be, at bottom, a physical phenomenon of any kind, for all phenomena—particles, waves, motion, extension in 'space', duration in 'time'—are to be *accounted for by* ether, but they cannot themselves *account for* ether. Any approach along those lines would soon be found to be tautological; it would lead to a circular argument which would leave us at the end where we were at the beginning. Nevertheless, some kind of inner differentiation there must be: otherwise we shall have nothing at the base of the cosmos but an ocean, so to speak, of something or other, which exists everywhere and always, but which is completely featureless: it could not therefore be defined or described, or be brought into relation with the highly diversified universe, which the realist philosopher accepts as fact, and takes as his starting-point.

4

I have ventured to suggest, in my previous book, a differentiation which would not infringe any of those limitations. It is not a differentiation between Here and There; nor between Now and Then. Nor would it be merely of a mathematical, or any other kind of symbolic order. Nor is it unfamiliar, or hard to understand. It has the cardinal virtue of simplicity. For it rests upon two physical characters with which everyone, scientist or layman, is well acquainted.

The subject-matter remaining the same, one differentiation proposed is between *states*; the other differentiation is between *patterns*.

Take as an illustration the most widely diffused and most familiar substance in the world of our perception — water. Chemically, every molecule of water consists of two atoms of hydrogen and one atom of oxygen, held together by binding electrical forces: it is H_2O. These molecules of H_2O may exist, however, at different times or places, in one or other of several different states. The commonest is the liquid: but there are also the gaseous state—steam (in volume 1,600 times as large as the liquid—hence the steam-engine), and the solid state—ice.

Secondly, the same molecules may exist in a variety of patterns. The liquid may become a vapour, diffused perhaps in the earth's atmosphere among the molecules or atoms of nitrogen, oxygen and carbonic-acid gas; or made visible as clouds. Or the vapour may be condensed into droplets of dew, or drops of rain, or, if driven by the wind from the crests of waves in the sea, may become momentarily hard stinging drops of spray. Or in the upper atmosphere, under certain conditions of pressure and temperature, the H_2O molecules may solidify into patterns of soft snow, or into hard hailstones of ice. Here we have a number of different states or patterns exhibited by a single substance, which itself remains unchanged.

Any one of these states or patterns may itself be transformed into one of the others if the environment is altered. We can watch the process going on, for example in a pond, whenever there is a hard frost: we see the flakes of solid ice forming—easily and quickly—in the liquid water. Or again, if a kettle is on the boil, the bubbles of gaseous steam appear at the bottom and press out violently through the spout. Any number of illustrations might be given. Indeed, chemists have found ways to transform most of the elements into the gaseous, the liquid or the solid state. Further, things may be hot or cold: the molecules of a piece of iron may be at one time in a state of greater thermal agitation and at another time of lesser, and a touch will tell us the difference. Or the air around us may have been yesterday windy and be today in a state of calm; or the sea rough or smooth.

It is often remarked that the same model is frequently being used in nature's engineering in different ways in different circumstances. The pattern of the wave, for instance, with its succession of crests and troughs, advancing across a two-dimensional surface or in three-dimensional spheres—'a moving configuration', as Lord Cherwell described it—is visible to us in the colours of the spectrum, or in ripples on a pond or waves in a sea; or audible in sounds in the air; or revealed by physicists in all the diverse phenomena of the electromagnetic wave-band.

Should an ether in fact exist, it would therefore not be surprising if its mechanism were to include differences of states or of patterns on similar models to those.

5

We may find it convenient for the purpose of our discussion to divide our problem of ether into two problems, needing separate answers: the first—what is the ether? The second—how does it work?

The answer to the first would be—energy. The answer suggested to the second (and this, so far as I am aware, would be a novel point) is that energy exists in either of two states—quiescent and active; that its essential characteristic is that it is able to pass easily from the quiescent to the active, and back again; that active energy may shape itself into different patterns; and that these mutations of states and patterns produce all the phenomena of the given physical universe.

On the question of quiescent energy there is little to be said. For the very reason that it is quiescent it can exhibit no phenomena. Universal and eternal, so long as it is in the quiescent state it is immobile and featureless; and therefore, incidentally, can give no hold to mathematics. It may be asked, why then complicate matters by bringing it in at all? The answer is—because quiescent energy is the matrix of active energy in all its forms, and therefore cannot be ignored by a realist cosmogony. Indeed, if the physicists of the next generation rescue the ether from its present low estate and give it recognition as the fundamental element in their science, it would follow that one of the chief objects of their study would be to discover the factors that cause the transmutations of quiescent energy into active and their relapse. Meantime I have nothing to suggest as to the events or conditions that are able to stir, at some particular time and place, the lethargy of quiescent energy into activity. One thing, however, seems clear: once the process of activation is set going it can transmit that activity to the neighbouring region of quiescent ether before relapsing into inactivity. This is the mechanism of the wave, in all its various manifestations: it is visible to us in the smooth pond when a stone is thrown into it—a first ripple is started around the hole in the water made by the stone, which sets going a second ripple, and so on with a third; and successively until the process dies away, or is stopped on reaching the solid bank. But with the ether we do not yet know what is the equivalent to the stone falling into the water.

6

This hypothesis—based upon the notions of energy and its states and patterns—might make possible a great simplification of the highly complicated picture presented by modern physics. It might offer simultaneous answers to some of the questions put by philosophers to scientists, which are now brushed aside. First, and most obvious, is the question what is the mechanism for the transmission of light and heat.

The idea of 'action at a distance'—some kind of emission from one place, followed by a reception at another after a jump from the one to the other, through a sheer vacuum, with no physical process of any sort going on in between—this is wholly unacceptable. It well deserves the scorn with which Newton treated it. Referring to gravitation—but the same must apply equally to all transmissions—he wrote, 'that one body may act upon another at a distance, through a vacuum, without the mediation of anything else by and through which their action may be conveyed from one to another, is to me so great an absurdity that I believe no man, who has in philosophical matters a competent faculty of thinking, can ever fall into it'. And Einstein says: 'As a result of the more careful study of electromagnetic phenomena, we have come (in modern physics) to regard action at a distance as a process impossible without the intervention of some intermediary medium'. He has written also: 'Physics has to represent a reality in space and time without phantom actions over distances'.

That a process is going on to carry, for example, the light of the Sun, and its heat, to the Earth is proved by our daily experience, whenever some opaque object intervenes. I may put up my hand to shade my eyes from the bright sunlight, or may open an umbrella; or a cloud may drift in between; or, during a solar eclipse, when the motion of the Moon brings it across the line between the Earth and the Sun—it is impossible to deny that in each case there is an interruption in the transmission, at the level of my hand or my umbrella, or at the level of the cloud, or of the Moon's orbit—such as was not occurring before that interruption, and will cease to occur when it is over. This is a constant experience, not only of all human beings, but also of animal and vegetable organisms in general. It follows that the idea of a jump across a vacuum is nonsense.

That refutation is conclusive. But in addition it may be observed that the transmission is a time process, the light and radiant heat travelling at the ascertained velocity of 186,000 miles a second—taking eight minutes from Sun to Earth, and thousands of millions of years from the distant galaxies. But if there were a jump, without any physical process going on between, it would necessarily be instantaneous, and we know that it is not.

'Action at a distance,' then, having been dismissed from the discussion, we have to ask what alternative there might be. If we seek it in some hypothesis or an ether consisting of energy in two states, quiescent and active, the next question must be what is the mechanism of the activation. In the case of the wave radiation of the Sun's light and heat—or any other kind of electromagnetic wave radiation—the first stimulus might perhaps be caused by some kind of vibration or oscillation at a point anywhere in the expanse of quiescent energy. The spherical layer of quiescent energy directly surrounding that point, would immediately itself become activated. This would in turn convey an activating impact on the next spherical layer, while the first one simultaneously relapsed into quiescence; and so on, in a succession of pulses, giving rise to waves with their crests and troughs. This may continue indefinitely, unless something is encountered that may stop it, or unless the originating cause ceases to operate;

or else until the expansion of the spheres, one beyond the other, diffuses the impulse more and more, until in the end it is spent.

This hypothesis involves changes in the present standard vocabulary of physics. It does not conceive that energy can be 'released', or 'transferred', or 'expended'; or, on the other hand, be 'conserved'. Nor do we suppose that any energy is travelling across from—for example—the place where the Sun is to the place where the Earth is: just as no volume of water travels across the surface of the pond from the place where the stone has fallen to the bank. In both cases what travels is the pattern, the 'moving configuration'. The quiescent energy in the one case, like the water in the other, remains where it is; it was there before, and will still be there afterwards. The motion is an up-and-down motion.

Let me add one other point which is relevant here, and which I have not included in my previous writings about ether. The use in nature of the same model for the mechanism of different processes in different circumstances may be seen in the similarity between the wave processes of ether radiation (if there is an ether) and the ripples on a pond when its surface has been disturbed. There must be some factual reason for such resemblances: it cannot be a mere chance or coincidence; and we are inclined to search for some kind of common origin for phenomena which are apparently distinct and diverse. But may there not be a different approach leading to a different explanation? May it not be identically the same phenomenon in the two cases—perhaps in more than two, but viewed at different size-levels?

In other words the ripples on the pond are themselves an ether phenomenon. What is happening is that the fall of the stone, under the influence of the Earth's gravity, has set up, in the quiescent ether, waves of energy activation in the shape of a train of spherical waves. But there are volumes of activation already present in the area in patterns of atoms and molecules—namely, the liquid H_2O. This is at a much higher size-level, but must necessarily take the same configuration. If, instead of water, the environment had been one of air, the formation would still have had to be one conforming to the same shape or pattern; we should then have had sound-waves in the atmosphere. (This may have a bearing on the dynamics of sound, the sound barrier and its mysterious 'bangs', the Mach measurement, and so forth.) Instead of saying how strange it is that the same model should reappear in quite different media, we should begin to talk about one phenomenon appearing in the same pattern but at different size-levels.

We may be surprised to find ourselves back again at those persistent 'tables', when we discussed the co-existence of quite different things—wood, cells, molecules, atoms, particles—at a succession of lower and lower size-levels, but keeping the same over-all shape.

All this ignores the question of the quantum, which lies far away at the remote size-level, relatively to our energy measurements, of 10^{-27} erg-secs.

7

The train of expanding spherical waves, then, is one pattern of activation. Another is the particle, with which we have become well acquainted during the present century through the widespread research into the structure of the atom, and especially of its nucleus.

Like the wave, it is very far away from us in size-level, in the region, relative to the standards of measurement of our terrestrial environment, of 10^{-14}. Particles differ from the wave-pulses in that they are discrete entities, which the wave-pulses are not; and they are not auto-mobile, as the wave-pulses are. Particles differ among themselves in pattern; a flow of information with regard to them is coming to us year by year from the vast atomic research establishments with which we are familiar in this country at Harwell, and others elsewhere.

There is also a third pattern of ether activation—the 'lines of force', of which the specific characteristic is that its manifestation is linear. Physicists do not tell us very much about lines of force; but Sir Edmund Whittaker discusses them, and mentions that Faraday 'constantly thought in terms of lines of force', especially in relation to magnetism. And Sir George Thomson tells me that they were one of his father's 'favourite ideas and that he did much work on them'.

PROFESSOR DINGLE:

1

I am glad that you have begun the discussion of physical problems with the origin of the relativity theory, for that is, I believe, the great crisis in modern physical history. What you say about physicists 'turning to mathematics and its symbolism to furnish a substitute' for a physical interpretation of experience is profoundly true. It has come about largely unconsciously. The relativity theory marks the final yielding to a tendency that began with Maxwell's theory of the electromagnetic field. Maxwell derived a set of equations, ostensibly by the recognized methods of physics—i.e. he postulated processes going on in electrified and magnetized bodies, in electric currents and in the surrounding ether, and he wrote down symbols to represent the measures of those things and equations showing the postulated relations between them. These equations did in fact represent observation with great accuracy and gave birth to new knowledge, the most striking example of which was that a certain relation between static electric charges and electric currents (expressed as the ratio of electrostatic and electromagnetic units) was equal to the velocity of light in empty space, i.e. in free ether. This established the view that light was an electromagnetic phenomenon, which had long been suspected.

This was eminently satisfactory, but all attempts to form a clear idea of what was going on in the physical systems concerned met with frustration. Maxwell's ideas proved curiously elusive: he said incompatible things in different places, yet both were necessary for the complete system. One of the greatest of his followers, Hertz, confessed that he could not follow the workings of Maxwell's mind, and concluded in exasperation: 'Maxwell's theory is — Maxwell's equations!' That was the beginning of the acceptance of mathematical formulae as a substitute for physical theory.

The history of the subject from this point is so important for our discussion and so remarkably little understood—or rather so greatly misunderstood—that I venture to recount it before proceeding to the very interesting suggestions you have advanced. For it is in this field of thought that the relativity theory was sown, took root, and grew up: the Michelson-Morley experiment is but one of many tests of electromagnetic theory, and although it is the simplest starting-point for a restricted account of Einstein's work, much of the wider significance of this phase of scientific history is lost if we regard it merely as an abortive attempt to detect the absolute motion of the Earth.

Maxwell's theory was directly applicable only to systems of bodies at rest in the ether. The first apparently satisfactory attempt to extend it to moving systems, and to take account also of the existence of particles of electricity (electrons), discovered after Maxwell's death, was made by Lorentz. He generalized Maxwell's theory—directed, unlike Maxwell, by the rational implications of plausible assumptions rather than by inscrutable intuition—and arrived at a set of equations which became universally accepted. I will call them the ML equations; all we need to know of them is that they are perfectly definite mathematical expressions in which each symbol represents, directly or indirectly, something which we can actually measure. They gave a remarkably accurate account of many experimental facts, and almost unlimited confidence was given to them.

But, surprisingly, they broke down when applied to bodies moving much faster than was then normal in physical laboratories, velocities such as those met with in astronomy. The Earth's orbital velocity is $18\frac{1}{2}$ miles a second — one ten-thousandth of the speed of light — and its motion is constantly changing direction. This should have produced easily observable effects, including that sought by Michelson and Morley, but none of them could be found. Everything happened as though the Earth were permanently at rest in the ether. No one could believe this, so an impasse was reached.

A partial escape was suggested by FitzGerald, but a complete resolution of the problem was first proposed by Lorentz himself. He made some assumptions, of which the chief were that an electric charge when moving through ether is contracted in the line of motion by a factor, which I will call α, depending on its velocity; that all rhythmical processes occurring on it are slowed down by the same factor α; and that these changes are suffered also by *uncharged* bodies. Assuming further that all mass is electromagnetic in origin, he then showed that a body should become more massive with motion, also by the factor α. This increase of mass (or, strictly speaking, of momentum) had, in fact, been observed with swift electrons, so there was some support for these ideas. Nevertheless, Lorentz realized their speculative character, and he remarked: 'It need hardly be said that the present theory is put forward with all due reserve'.

Two things should be said about it. First, Lorentz's proposals were not modifications of his theory but additions to it. The theory was complete in itself but it led to the wrong results. When supplemented by these independent assumptions it gave the right results, but no other reason could be found for the assumptions. Secondly, the assumptions emphasized what was already an essential part of the theory, that all phenomena occurred in a static ether penetrating all bodies, even electrons, which was unaffected by anything that matter could do. The postulated contraction, slowing down, etc., were real, not ideal, physical changes. 'There can be no question about the reality of this change of length,' Lorentz wrote so late as 1921; a moving rod was actually shortened 'just as it would be if it were kept at a lower temperature'. That meant that the ether must be as real as the rod, for without this standard of rest a moving rod cannot be distinguished from a stationary one.

With these assumptions Lorentz derived equations which gave the relation between space and time measurements in stationary and moving systems. These equations, known as the *Lorentz transformation equations*, are as important for our purpose as the ML equations of electromagnetism. Again we need not know what they are; I will call them the LT equations. They supplemented the ML equations, and the two together enabled a sufficiently good mathematician to solve all electromagnetic and optical problems in stationary and moving systems. What it all came to was this. In a stationary system the ML equations were sufficient. In a moving system you applied the ML equations and 'corrected' the result by the LT equations. It then followed that all the effects of the Earth's motion, that had been expected but had failed to appear, were no longer to be expected; the LT equations wiped them out. It followed also that it would be for ever impossible to determine, from observations within a material system, whether the system was resting or moving uniformly through the ether; such observations would be the same in both cases. This was already known for mechanical phenomena, so it became the conviction of physicists that no observation of any kind at all, made within a material system, could reveal the state of motion or rest of that system.[1]

But, as I have said, there was nevertheless a real difference between rest and motion, and between motion with one velocity and motion with another: the reality of the ether ensured that. And, if Lorentz's ideas were right, this could be determined by comparing two bodies in relative motion. Suppose, for instance,

---

[1] In view of what is to follow, it should be noted that this, and this only, was what Poincaré called 'the principle of relativity'.

that side by side there are two identical clocks showing the same time and going at the same rate. Now suppose one moves rapidly away and returns. Suppose for simplicity that they were initially at rest in the ether; although we cannot know this, the result to be described will be the same, except for small differences, whatever their initial motion might be. The travelling clock will then run slow while moving, so on return it will show an earlier time than the other. This applies to all rhythmical processes, including those of our bodies if we suppose them to obey the ordinary physical laws. Hence a returned space-traveller will have aged less than ourselves. And if, as we may suppose, his psychological experiences will be correlated with his physiological processes in the same manner as ours, he will also seem to himself to have lived a shorter time. In an extreme case, if he moves fast enough he may come back after what seems to him only a few days, to find his acquaintances long dead and their descendants alive.

This consequence of the theory has received much attention since, but none, so far as I know, at the time. It follows from the physical reality of Lorentz's proposed changes, and therefore of the ether and of motion with respect to it. The next step was taken in 1905, when Einstein published his basic paper on what is now called the *special theory of relativity*. Starting *ab initio*, he made two postulates, the 'postulate of relativity' and the 'postulate of constant light velocity'. The former asserted that 'the phenomena of electrodynamics as well as of mechanics possess no properties corresponding to the idea of absolute rest'. This had long been accepted in mechanics; Einstein's postulate extended it to electromagnetism also.

This was a direct denial of Lorentz's essential requirement, a stationary ether. Einstein saw this, and stated that on his theory 'the introduction of a "luminiferous ether" will prove to be superfluous'. It did not follow, of course, that space might not contain something with physical properties which might as well be called ether as anything else, provided that it could not serve as a standard by which a particular velocity could be assigned to a body. To take an example, Faraday[1] once suggested that every atom carried with it a system of 'rays' or 'lines of force', extending indefinitely in all directions and moving instantaneously with it. The rays from one atom could freely interpenetrate those from others, and they could serve as a vehicle for transmitting

[1] *Philosophical Magazine*, 28, 345 (1846).

light and electromagnetic influences. The ensemble of all such rays in space can be called an ether, but it would be impossible to ascribe to any body a velocity with respect to it. The body must always be at rest with respect to its own rays, and have various velocities with respect to all other sets of relatively moving rays, so that the only measurable velocity would be that of one body with respect to another.

I shall return to this idea (which, I should make clear, was not invoked by Einstein in this connection), but I cite it now only to show the possibility of an ether that does not impose a unique velocity on a body and so violate Einstein's postulate of relativity. But Lorentz's ether essentially did impose a particular velocity, which in principle could be established by the contraction, slowing down, etc., described above. Hence, at its very first step, Einstein's new theory departed irreconcilably from the theory of Lorentz.

But its next step took it back into the Lorentz theory again. Einstein's second postulate said that 'light is always propagated in empty space with a definite velocity c which is independent of the state of motion of the emitting body'. That means that if you have two bodies in relative motion, and at the instant at which they pass one another each emits a beam of light in the line of motion, these two beams will travel through space together as one beam, and reach a distant point simultaneously. Now this is just what would happen if light travelled as waves in a stationary medium. Once started, their velocity, no matter how their *source* was moving, would be fixed by the properties of the medium, just as the velocity of sound waves in air depends only on the state of the air. On the old corpuscular theory of light, however, each beam would travel at velocity c with respect to its source, and so the beam from the body moving in the direction of the light would reach the distant point first. This postulate therefore gave back to Lorentz much of what the first had taken away. There was no Lorentz ether, but light travelled through space as though there were.

Unfortunately, an experimental test of this postulate was then impossible, and the postulate had to remain an assumption. I very much hope that the experiment can soon be made, and then much of our present uncertainty can be removed. In the meantime, however, we must be content with testing what we can infer logically from it.

Einstein's *tour de force*, however—and, quite apart from the question of its validity, as an intellectual achievement it compels our highest admiration — was that he reconciled his two apparently contradictory postulates and derived the LT equations— which, you will remember, were in Lorentz's system an *ad hoc* supplement to the ML equations—as a necessary consequence of our methods of measuring space and time. Whatever our state of motion, we were to synchronize distant clocks in the same state of motion by assuming that light, having the properties given it by the second postulate, took equal times to travel to and from any such clock; and further, that, having once been so synchronized, clocks in good working order would remain so. It then followed by pure logic that the relations between measurements in relatively moving systems would be those given by the LT equations. The ML equations Einstein took for granted, so that, mathematically, his system was identical with that of Lorentz. It had the overwhelming advantage, however, that it gave a *reason* for the LT equations, and so for their implications; they were consequences of the fundamental postulates, not *ad hoc* additions.

But although identical mathematically, physically the two systems were quite incompatible. In the Lorentz system, as we have seen, the real slowing down of a clock moving through the real ether would appear when the clock returned. But if there were no ether, the 'moving' clock could not be distinguished, so no retardation could appear. What, then, was the physical meaning on Einstein's theory of the 'slowing down' which the equations indicated? The answer is that when the clocks are separated you can compare their rates only by using light or similar signals to ascertain the reading of one from the position of the other. It then followed that (light having the postulated properties, and separated clocks being synchronized in the assigned way) an observer with *each* clock would find that the other's time for any event was behind his own. It would be like two men, each 6 feet high, separated and asked to estimate their relative sizes. They could not use the same tape-measure, so each would have to measure the apparent, the angular, size of the other, and each would find that the more distant the other, the smaller he would appear. There would be no real shrinking with distance, and on reunion their heights would be found equal. Substitute motion for separation and times for heights, and you have what Einstein's theory requires of returning space-travellers.

This is all, I think, perfectly clear, but now begins the woeful tale of misunderstanding. Contrary to common belief, Einstein's theory did not for many years attract much attention. Since it had the same *mathematical* content as Lorentz's theory, it was taken to be another form of that theory and Lorentz, who clearly had priority, was taken as its author. Ritz, in 1908, in opposing the theory, scarcely mentions Einstein; it is Lorentz whom he attacks. And Poincaré, so late as 1912, speaks of 'the relativity theory of Lorentz', ignoring Einstein entirely. And yet the *relativity* part of the theory was Einstein's alone.

How could such a thing happen? On examination it becomes clear that what Poincaré means by 'relativity' is not Einstein's postulate that nature knows no absolute standard of rest, but the much older postulate that the state of motion of a system cannot be found from observations *within the system*. With this Lorentz agreed; he invented the LT equations to satisfy it. But Einstein's relativity—that the Lorentz ether was superfluous and the only meaningful motion was that of one body relative to another—was directly denied by the theory of Lorentz. Thus began a confusion that still bedevils discussion of this subject; the same word, 'relativity', is used for quite distinct ideas.

That is not all. Quite apart from the name, essential physical requirements of Lorentz's and Einstein's theories, which are flatly incompatible, are now almost universally thought to be identical. Since Lorentz's theory came first, and Einstein's was for long thought to be merely another form of it, it is the Lorentz physical requirements that are usually accepted, but ascribed to Einstein. In fairness it must be recorded that Einstein himself sometimes made the same confusion. But the facts are inescapable, and only those ignorant of history can equate greatness with freedom from error. Genius is almost invariably accompanied by liability to elementary mistakes, and the authority of great men no longer *legitimately* carries weight in science.

We have here, then, the plainest confirmation of your charge that mathematics has become a substitute for physical explanations. Mathematically the Lorentz and Einstein theories are identical; physically they are irreconcilable. And almost universally the mathematics is regarded as the essence of the matter and held to justify the holding of both theories together. 'The special theory of relativity', a world-famous physicist wrote me recently, 'so far as it is specified and defined by the transformation equations of Lorentz (and this, to my mind, is the only proper way of specifying what the special theory is), is a completely self-consistent edifice.' So presumably Einstein contributed nothing. Yet Lorentz, in 1928, said: 'the theory of relativity is really solely Einstein's work'.[1]

And indeed, nothing could show more clearly that the currently accepted theory is that of Lorentz, falsely attributed to Einstein, than the late Sir Edmund Whittaker's recent 'History' to which you have referred in Appendix 1. Whittaker describes what we call 'the special theory of relativity' in a chapter entitled, 'The Relativity Theory of Poincaré and Lorentz'; he nowhere mentions a 'Relativity Theory of Einstein'. This has caused indignation among the disciples of Einstein, but if, with them and Whittaker, we regard the mathematics as the essence of the theory, Whittaker is clearly right. But, unlike most of them, he knew his history, and he knew that the mathematics of the theory was wholly Lorentz's. But Lorentz never regarded it as a *relativity* theory; he knew perfectly well that his theory was *not* a relativity theory because it was founded on an absolute ether. But it had an aspect to which the name 'relativity' could be attached (see p. 64) and this aspect was emphasized by Poincaré. Accordingly, Whittaker had to include Poincaré with Lorentz in order to retain the name now universally given to the theory in question. That is the whole story, and, if the theory is defined by its mathematics, Whittaker is irreproachable: there is only one theory, and it belongs to Lorentz. But if the theory is to be defined by *the physical ideas expressed by the mathematics*, there are two distinct theories, that of Lorentz and that of Einstein. The former cannot exist without a nineteenth-century stagnant ether; the latter cannot exist with it. The difference is fundamental. And almost the entire world of mathematical physicists, gone a whoring after empty symbols, has become blind to this difference; it takes the ideas of Lorentz and gives them the name of Einstein.

2

I wish the story could stop there, in which case your plea for a return to scientific principles might evoke a response; but

[1] *Astrophysical Journal*, 68, 350, 1928.

unfortunately that is not so. This trafficking with arbitrary abstractions has not only led scientists astray; it has proved so seductive that it has paralyzed their powers of return, and nothing but a shock treatment such as you have initiated (or some physical disaster which it is more comfortable not to contemplate) offers any prospect of recovery. It is necessary to relate a little contemporary history, which may seem incredible: I can vouch for its truth, and I do not think its gravity needs emphasis.

For the last five years I have been engaged in a continuous controversy—partly in published papers and much more extensively in private correspondence—with physicists and mathematicians of all degrees of eminence, all over the world, on the problem (pp. 65-8) of the space-traveller, which in principle is the simplest means of distinguishing between the Lorentz and Einstein theories. Almost to a man they maintain that Einstein's postulate of relativity is true, but nevertheless the returning traveller will be abnormally young, as Lorentz's theory requires. The impossibility of this is obvious to an intelligent schoolboy, and after a protracted bombardment by spurious and mutually contradictory attempts by the world's leading scientists to make it feasible, you can perhaps imagine that your sane physical approach to such problems comes as a balm. Since high-speed space-travel might become a reality, the idea that it is possible to postpone death on the terrestrial scale by a few hundred years (which continues to be preached in the press all over the world as a scientific fact) may have momentous implications not only for our thinking but for our political and other actions also. We are dealing not with a scientific curiosity but with the most practical affairs.

The simple facts are these. When the traveller sets out, there is nothing, in the absence of a Lorentz ether, to affix the (relative) motion to him rather than to the Earth. Hence we cannot say that his clock rather than our clock will lag behind, so when he returns there is nothing (except, of course, the possible direct effects of any applied forces, which will vary with the forces and are agreed to be irrelevant) to cause a difference between the clock readings, which must therefore agree. Our periodic bodily processes are equivalent to clocks for this purpose, so the traveller will age at the same rate as ourselves.

In this argument no flaw has been indicated: I have, however, received a host of independent incompatible 'proofs' that the traveller will return younger by precisely the amount that the Lorentz theory postulates—all from disciples of Einstein. Here are three.

The first says simply that the equations require it. The co-ordinates in the equations represent the times which the clocks will show. You apply the equations, and the co-ordinates demand that the traveller's clock shall show the earlier time. That settles the matter. The fact that the traveller has to reverse his motion is unimportant. According to the second account, it is true that if you consider the outward and inward journeys alone, you cannot distinguish one clock from the other; but the traveller's clock must reverse its motion, and so must be *accelerated*. To deal with this you must bring in the *general* theory of relativity. This shows that the effect of acceleration alters the times so as to give precisely the same retardation of the traveller's clock, no matter which you suppose to have moved. The third treatment says, like the second, that you cannot solve the problem without taking the acceleration into account, but, unlike the second, it regards this as making it insoluble. The alleged proofs are 'sham proofs'; that is all.

Now clearly, if any one of these arguments is right, the others must be wrong. There are many other solutions, showing still more incompatibilities, but I cite these three because they are all from Nobel Prizemen in physics. We are dealing not with the blundering of students but with the considered utterances of the leaders of scientific thought. One would have expected our philosophers—particularly those of the logical positivist type, who hold the function of philosophy to be the criticism of scientific statements—to have rushed in to restore reason, but they are unanimously silent. They seem preoccupied with analyzing the various ways in which banal statements could, but never would, be misunderstood if anyone were puerile enough to utter them. Of real criticism of science they have nothing to offer. I will revert to this presently.

And this, bad as it is, is not the worst: the sense of responsibility seems to have suffered the same paralysis as the reasoning faculty. In the early stages of the controversy I believed that Einstein's theory, when properly interpreted, was true, and I differed from the others only on its interpretation. But in analyzing the various arguments for this 'asymmetrical ageing', as it is called (my opponents did not do me the honour of analyzing my simple statement; they consistently withheld comment), I came to see that Einstein's theory could not be true (which does not mean, of course, that Lorentz's necessarily is). I saw no reason to question the first postulate, which prohibits asymmetrical ageing through motion, but I was able to show, I think conclusively, that the second, coupled with the chosen method of synchronizing separated clocks, led to contradictions. I also saw what I believe to be the necessary amendment, but that is not on the same level of certainty as the inadmissibility of the present form of the theory, which I think is a logical necessity.

I put my criticism and the proposed amendment into a paper which was submitted to the Royal Society. A referee reported that it was 'a smother of words' (it is true that it was less besprinkled with symbols than most such papers), that its criticisms of Einstein's theory were 'palpably false', and that the proposed amendment was made to meet 'anomalies that do not exist'. That exhausted his comment on the criticism and proposals, and the rest of the report concerned minutiae. Without sharing the referee's views on these, I cut them out and asked if the paper might be submitted to someone who would show me where the 'palpable falsity' lay. I was not so enlightened, but was informed that a second referee had recommended rejection, and that course was adopted.

Now the Royal Society has a wise rule for referees which says: 'A paper should not be recommended for rejection merely because the referee disagrees with the opinions or conclusions it contains, unless fallacious reasoning or experimental error is unmistakably evident.' Since in my paper neither fault was evident to me, and the referees had not attempted to make it so, I wrote to the Royal Society, asking for information on this point. My letter elicited an acknowledgment, but no other response.

I next sent the paper to *The Philosophical Magazine*. As you know, this journal originated at the end of the eighteenth century, when the present sharp difference between the terms 'science' and 'philosophy' did not exist; it has always been a journal devoted to physical science, which was once recognized as a part of philosophy. My paper, however, came back by return of post with the statement that 'subjects of a polemical nature are not suited to [this] journal'. You will at once, and quite rightly, see here an example of the dangerous separation between science and philosophy that led you to originate this discussion, but there is another and equally menacing aspect of this remark. One of the leading scientific journals will not publish anything 'of a polemical nature', which can only mean that, in science itself, it will not publish any criticism of orthodox views. Accept them, and your paper will be considered for publication; question them, and it will not. This is a *philosophical* magazine—the magazine in which Faraday's 'thoughts on ray-vibrations' (p. 65) are preserved for us.

However, the paper was next submitted to the Physical Society by a friend who made a special request that, if it were not published, the reason might be given, as it seemed to him quite sound. It was kept for five months and then returned with a statement that it had received the most careful consideration but the Society was not prepared to publish it. No criticism of it was offered, no reason for rejection was given, and no reference was made to the request for a reason.

In the meantime I had been able to locate the internal contradiction of Einstein's theory in a single point and to present it in a form in which I think it is no longer possible to question it. Since it involves next to no mathematics, and is easily comprehensible to a sixth form schoolboy, I venture to give it in an Appendix (Appendix 2). This I sent to *Nature* for publication. It was referred to a critic 'particularly competent in the field of relativity', who after several weeks reported on it. He worked out in detail the formula which I had accredited to Einstein's theory, and found it correct. He also correctly located the point at which my argument deviated from Einstein, and he went on: 'It seems to me that the deviation takes place at the assertion referred to above. One need read no further. Why does Professor Dingle make this assertion? I do not know . . . I can only conclude that he has a private way of looking at these matters which does not agree with the general way.'

You will see that my statement is throughout a piece of overt reasoning, with nothing resembling 'a private way' about it. If there is a fallacy it can be precisely located. So what the critic says is this: I can see nothing wrong with this refutation of Einstein's theory, but it doesn't agree with the theory, so one need read no further.

The Editor of *Nature* acted as 'postman' in further discussion with the referee, which brought nothing but paraphrases of his previous remarks. Ultimately the Editor decided that 'It is quite clear from the correspondence that neither you nor he can appreciate the position taken up by the other. In the circumstances I cannot see that any useful purpose would be served by opening a discussion of the subject in *Nature.*' On this I will say only that whether or not I understood the referee (I have no doubt that I did) is immaterial since he had submitted nothing for publication. But if he failed to understand me, then either my statement was obscure or else he was thereby disqualified from judging it. You may decide between these alternatives.

I will not recount the dreary list of similar rejections, in other countries as well as here. They have this in common—a refusal to give any reason for rejection or any criticism of the arguments advanced, except that the conclusion is opposed to current theory.[1]

'Science,' wrote T. H. Huxley, 'seems to me to teach in the highest and strongest manner the great truth which is embodied in the Christian conception of entire surrender to the will of God. Sit down before fact as a little child, be prepared to give up every pre-conceived notion, follow humbly wherever and to whatsoever abysses Nature leads, or you shall learn nothing. I have only begun to learn content and peace of mind since I have resolved, at all risks, to do this.'

For long I believed that this could have been written with truth by most scientists, both individually and through their organizations. Through sheer weight of evidence I can believe that no longer. I do not imagine that they are consciously traitors to their responsibilities: they have simply lost the power of understanding what they are doing. They have substituted mathematics for reasoning, and now automatically believe that its categorical character absolves them from considering any unorthodoxy.

We are concerned here only with matters of truth and error, not with their practical consequences, but I should not omit to mention what I have not failed to try to impress on the publishing agencies. If what I think is correct, workers in atomic energy establishments are systematically underestimating the velocities of the particles they are accelerating. I do not know what they

---

[1] Since this was written I have given a still simpler form of the same argument, which has been published in *Philosophy of Science* (Vol. 27, p. 233, 1960)—a journal which mathematical physicists do not read.

are doing, but I know that a situation in which one is using an instrument which he misunderstands is a dangerous one, and I know the kind of danger involved in these experiments.

LORD SAMUEL:

1

I have only one point to raise on your last piece, much of which is outside my range.

As to the two clocks, one 'in motion' and the other not, if these are theoretical clocks and part of a mathematical picture I have no comment to make. The theoretical observer may make observations with theoretical clocks and may reach whatever mathematical conclusion he chooses, according to the assumptions that he has made at the outset. But if, as seems to be the case here, a transition is attempted from abstractions to realities, and real clocks, such as you can buy in a shop, are to be employed in a practical experiment, a few questions may be allowed from the realist standpoint.

If you have those 'two identical clocks showing the same time and going at the same rate', and 'one moves rapidly away and returns', what is meant by saying that the moving clock is retarded? What is supposed to be *the physical process which brings this about*?

This seems to me not to be a matter for Einstein or Lorentz, Maxwell or Poincaré, or for mathematics at all, but for the designers and makers of clocks, for horologists. A clock is physically a closed system—subject no doubt to possible influences from the outside environment due to changes in atmospheric temperature or pressure, or the like. The watch that I carry in my pocket may have been keeping exactly correct time for weeks; it is quite unaffected by what is happening to it, so long as it is kept wound and is not stopped by being dropped or jolted. I do not suppose that any mathematical physicist would expect its hands to move slower, even infinitesimally, if I were to lend it to an air-pilot who was flying a jet plane round the globe in one or two days. Would timepieces in a Russian sputnik, at the end of half-a-dozen circuits round the Earth at very high velocities, show an earlier hour when compared with the Greenwich (or Moscow) mean time shown by clocks that had meanwhile been stationary? If the answer to that question is to be Yes—then I would ask: what has been the physical process by which the clock that has been in motion has been retarded?

The essential feature of an ordinary clock is a mainspring, which is coiled up tightly when the clock is wound, and which gradually releases itself after the clock has been set going. The release is made gradual by the attachment of the spring to a mechanism consisting of several inter-geared cogwheels, which finally moves the two hands round the dial, marked by divisions corresponding to twelve—or twenty-four—hours in the day. If the clock is found to be running fast, that can be corrected by someone altering the regulator that forms part of the mechanism, and similarly if it is running slow. All this is a matter of the relation between the mainspring and the hands, any change in this relation being brought about by human intervention. But with this real clock in a real space-ship—what is the process there? For one of the travellers to move the hands occasionally, or give a touch to the regulator, would clearly be inadmissible. What is the alternative? You will know whether the space-scientists have offered an answer.

Similarly with the question of the alleged effect on the space-travellers of suspending the normal process of physical ageing. You state the proposition in these terms—The instance of the clocks applies 'to all rhythmical processes, including those of our bodies if we suppose them to obey the ordinary physical laws. Hence a returned space-traveller will have aged less than ourselves. And if, as we may suppose, his psychological experiences will be correlated with his physiological processes in the same manner as ours, he will also seem to himself to have lived a shorter time. In an extreme case, if he moves fast enough he may come back after what seems to him only a few days, to find his acquaintances long dead and their descendants alive.'

My question here is similar. From my standpoint the problem is not mathematical but physiological. You set out (p. 71) the answers offered by three of the scientists who have written to you in defence of the postulate that 'the traveller will return younger by precisely the amount that the Lorentz theory postulates'; and you give an answer in terms of their own proposition. As to this I would not presume to offer any opinion: but I feel that any theory which is in such flagrant contradiction with common-sense would need much more powerful arguments before it would be likely to command any measure of general support.

2

Evidently you feel deeply on this matter. You are distressed that men who hold positions of leadership in the world of science should be so perverse as to hold and insist upon opinions such as these, and you even look forward to the possibility of devising a practical experiment which would test whether the Lorentz contraction can actually occur in nature. I confess that I cannot feel any such emotion. It seems to me that the hypothesis we are discussing is only one more of those amusing philosophic paradoxes which are put forward from time to time to test the ingenuity of junior students. It is on a par with Achilles and the tortoise, Buridan's ass, Epaminondas the Cretan, or the clever fellow who says 'I defy you to prove that two and two make four and not five'. In each case, of course, there is an assumption concealed in the wording of the question, which is tacitly ignored, and which is a false assumption. As soon as some form of practical test is applied the paradox is exposed and disappears.

In this case of the juvenility of space-travellers, by all means let a practical test be imagined, but if it is to test realities let it be physiological and not mathematical. And would it not be better—more feasible and more controllable—if the experiment were not made with an elderly human being, needing a long period for its testing and subject to many obvious difficulties? We might experiment equally well with some other animal organism, and with youth and growth rather than with age and decrepitude. Suppose for example that the laboratory cat has given birth to four kittens: it is intended at first to keep two and to drown the other two, but it is afterwards decided to reprieve those and to send them, as soon as they are weaned—about six weeks later—in a space-ship which is about to start on a voyage to some planet and which would take about twelve months for the double journey. During that time the kittens left at home will have grown into adult cats, and perhaps have already engendered offspring of their own; while their brothers, the travellers, would have aged hardly at all and would retain all the engaging qualities of kittens less than six months old.

The only question I would ask would be—not whether this remarkable phenomenon corresponded with the equations of Lorentz or with those of Einstein. My question would be whether the experimental animals had been fed at the time intervals to which they had been accustomed before they had been weaned:

or whether hunger and thirst, alimentation, digestion, absorption or excretion, had been spread over much longer time intervals in the rhythm indicated on the retarded clock on the spaceship. If the former, the experiment would not have been a test of the problem. If the latter, the unfortunate animals, their protests having been disregarded, would in all probability have died of starvation long before. The case might even engage the attention of the Royal Society for the Prevention of Cruelty to Animals.

I suspect that the tacit assumption, in this problem of the juvenility of space-travellers, lies in the notion that 'Time' is *something*—that it is a factor present in the given universe which must be taken into account by philosophers and mathematical physicists, and be represented by a symbol, 'T', and particular 'times' by other symbols '$t_1, t_2, t_3 \ldots$'. With this goes the 'retardation' or 'acceleration', and the 'velocity', of this element 'Time'. This assumption being false, the consequence is that, when the endeavour is made to re-translate those abstractions back into realities the result is a barren paradox.

PROFESSOR DINGLE:

1

Your reaction to my account of this controversy is that to be expected of any intelligent person whose reasoning power has not been destroyed or paralyzed by over-indulgence in symbol-manipulation: it is that of incredulity. You say first that 'I do not suppose that any mathematical physicist would expect [my watch's] hands to move slower, even infinitesimally, if I were to lend it to an air-pilot who was flying a jet-plane round the globe in one or two days'. And secondly, you say that if indeed such a thing should happen, you would ask 'what has been the physical process by which the clock that has been in motion has been retarded?' On the first point you are mistaken: almost all of them, no matter how eminent, do expect it. On the second point I cannot get them to give a direct answer but only an implied one, which is that the physical process is the *motion* which the clock possesses, notwithstanding that, as they agree, that motion is possessed no more by the clock in the aeroplane than by that which remains on the ground.

Let me try to convince you that this is the actual situation.

Professor W. H. McCrea, F.R.S., has written[1]: 'Were a new Isaac Newton born today, we could send him space-travelling so as to return to us in thirty years time at, say, the age of three. This would be in accordance with the theory of relativity and with experimental tests of the theory.' The implications of this 'fact' were then seriously imparted to the world. Notwithstanding its stark inaccuracy—no such experimental tests have ever been made—this nonsense was published in the leading scientific journal.

About the same time[2] I put the argument against the deduction of asymmetrical ageing from relativity into the form of a simple syllogism, so that the fallacy, if any, in it could be unmistakably indicated, and I invited anyone to tell me which step was false. Here is the syllogism:

1. According to relativity, if two bodies (e.g. two identical clocks) separate and reunite, there is no observable phenomenon that will enable one to say, in an absolute sense, that one rather than the other has moved.

2. If, on reunion, one clock is retarded by a quantity depending on the motion and the other is not, that phenomenon would enable one to say absolutely that the first had moved and not the second.

3. Hence, if relativity is true, the clocks must be retarded equally or not at all: in either case their readings will agree on reunion if they agreed at separation.

Professor McCrea replied (his reply follows my letter in the same issue of *Discovery* as follows: 'In Professor Dingle's letter, his statement (1) is demonstrably false . . . Of course, it is not necessary to say that "one rather than the other has moved" '. In other words: (1) is demonstrably false, but of course that does not mean that it is not true.

In August 1957, the BBC broadcast an exposition of the same supposed phenomenon by Professor H. Bondi, F.R.S. It is not in print, but an amplified form of it was published shortly afterwards.[3] The argument was that because a man going from London to York and thence to Bristol would travel further than one going direct from London to Bristol who left and arrived simultaneously with the other, therefore he would age less than the

---

[1] *Nature*, 179, 909 (1957).
[2] *Discovery*, XVIII, 174 (1957); *Nature*, 179, 1242 (1957); *Science*, 127, 158 (1958).
[3] *Discovery*, XVIII, 505 (1957).

other during the journey. If you do not see how the conclusion follows from the premises, I cannot help you: that was not explained.

But do not be depressed: the reason may be not that you are weak of intellect but simply that you are not a physicist. Professor Max Born—I need not remind you of his eminence in this subject—was so impressed by Bondi's published exposition that he recounted it at a Conference of the Evangelistic Academy of Loccum in Germany. To the account of his remarks,[1] however, he adds this pathetic footnote: 'Ich hatte den Eindruck, dass auch diese so einfache Überlegung von den in Loccum anwesenden Nicht-Physikern (Theologen, Philosophen, Juristen u.a.) nicht recht verstanden oder ihre Schönheit gewürdigt wurde'. (I had the impression that even this simple argument was not properly understood by the non-physicists—theologians, philosophers, lawyers and others—at Loccum, or its beauty appreciated).

I confess that my sympathies are with the non-physicists, but I understand Born's disappointment too. For when I read his reply to my syllogism I had the impression that even that simple argument was not properly understood or its beauty appreciated. His sole comment was this: 'Dieses Missverständnis zeigt die suggestive Wirkung des Wortes Relativitätstheorie'.[2] (This misunderstanding shows the force of suggestion residing in the word 'relativity'.) I may be wrong, but I have the feeling that I should have been more successful with the theologians, philosophers, lawyers and others.

I need hardly say that Born then proceeded to demonstrate the force of suggestion residing in mathematical symbols by proving through them that the impossible must nevertheless happen. It was not difficult to show that this 'proof' was fallacious, and I did this in a paper which I requested the distinguished communicator of Born and Biem's paper to submit to the Amsterdam Academy. He refused to do so. He made no attempt to find a fault in my paper, but simply said: 'I am not willing to lend a hand in promulgating a mist of misunderstanding on a problem which by now has been adequately treated and fully elucidated'. So the impression is left that Born and Biem's argument (which, incidentally, is irreconcilable with McCrea's, although it reaches

[1] *Physikalische Blätter*, 14 Jahrgang, Heft 5, p. 207 (1958).
[2] M. Born & W. Biem, *Proc. Amsterdam Acad.*, B61, No. 2, 112 (1958).

the same conclusion) is unanswerable, when in fact an answer is not permitted.

I could continue *ad nauseam*, but this will probably suffice. My syllogism is still unanswered—except by such answers as those given above; its conclusion is just dogmatically denied. So you may be assured that mathematical physicists of high repute *do* indeed assert, not merely that the air-pilot's watch would run slow infinitesimally, but that, at the speeds which may be attained in the not distant future, 'while centuries passed by travellers would experience only the passage of minutes and would become "almost immortal" '.[1] Your analogy with Buridan's ass and such paradoxes is thus not valid. No one, not Buridan himself, has ever believed that the ass would starve in the circumstances imagined, though difficulty might have been experienced in justifying such scepticism. But here the arrest of ageing *is* believed: there is no scepticism to be justified.

Furthermore, the public has learned to trust scientists' assertions, no matter how extravagant they may appear. In the political world, which you have done so much to redeem, however sordid in some respects it may be compared with the world of science, the saving grace of criticism has always been available. If a statesman claims to have discovered an infallible preventive of unemployment, only the most gullible believe him. But a scientist can utter almost any apparent absurdity, and it is swallowed without question. And indeed, in face of the many achievements that would once have been thought miraculous, who can be surprised at this? Asymmetrical ageing is not in itself more astonishing than television or photographing the far side of the Moon: why should not the ordinary person believe that he can postpone the date of death indefinitely, in view of the authoritative pronouncements I have mentioned? And what may not be the social consequences of such a belief?

Your amusing experiment with the kittens would be very relevant if it could be carried out, and I do not think the RSPCA would have cause for complaint, because the animals could be fed whenever they felt hungry without invalidating the test. Those who believe in asymmetrical ageing would not consider that physiology had anything to do with the matter, and the test would be equally valid with any projectile, organic or inorganic, that exhibited a rhythmical process. When it does become

[1] *The Times*, September 19th, 1956.

possible to make a direct experiment on the matter I think it most unlikely that living creatures will be used because the experimental difficulties and uncertainties are so great; the 'clock' will be an inorganic one of some kind. Your suspicion that the tacit assumption behind the expectation is that time is 'something' is, I think, very near the truth but is not quite right. Some go so far, in fact, as to claim that *symmetrical* ageing would imply the objectivity of time.[1] I think this is a complete mistake, but the actual assumption which they tacitly make is that not time but *motion*, as an intrinsic property of a single body, is 'something', notwithstanding that they deny this verbally. And, moreover, this motion has wonderful properties. If one pair of kittens has it, it makes them age slowly; if the other pair has it, it makes them age quickly. But how it is able to tell which is the actual case I am afraid I cannot tell you. Perhaps the kittens must be mathematicians to qualify for the experiment—but this is only a guess.

But in truth it is no laughing matter, for, much more important than the implications of this particular delusion, is the condition of which it is a symptom. It shows beyond question that science has lost its anchorage in the basic realities of experience and reason. I will try to find the cause of this in the next section of our discussion. Here I will only say that the question which, to you and to any rational being, seems an obvious one—what retards the traveller's physiological processes?—receives only this implicit answer: It is motion: true, the motion belongs no more to him than to the stay-at-home, but of course it does not affect the stay-at-home's physiological processes because he is not the traveller.

I can understand that all this seems incredible, but the inescapable fact is that, incredible or not, it is true. And the state of affairs which it reveals is unimaginably portentous. Science has now achieved such control that the future of life on this planet is in its hands. And when the men in whose hands the future of science lies — men who, by virtue of the implicit confidence which the public reposes in them, and of their own statutes, written or unwritten, should be dedicated to truth at all costs—can so forget their obligation as to meet criticism by suppression instead of reason, and persist in staking the existence of the human race on an assumption which they refuse to allow to be

[1] See, for instance, McCrea, World Science Review, April 1957.

questioned, then it becomes a duty to make these facts known in their plain nakedness, in order to restore the lost sense of responsibility, both to the truth and to mankind, before it is too late.

2

I expect the thought has occurred to you: 'Surely this is a matter on which our logicians should be able to help us. The syllogism which you have constructed (p. 79) is a very simple example of the traditional form of reasoning, and the content of the premisses involves only such very general notions that even an elementary student of philosophy should be able to decide on its validity without risk of error. What have our philosophers got to say about it?'

Alas, nothing could more emphatically justify your charge that 'philosophy has gone its own way, with little regard to the revolutionary conceptions of modern science' (p. 13) than the fact that no philosopher or logician has, to my knowledge, said a single word about the matter or shown the slightest awareness that the problem exists. 'Philosophy,' it seems, is quite other than this: its concern is with language, not with the ideas, however important, which language is used to express. While this controversy has been going on, a series of competitions has been running in the journal *Analysis*, in which problems have been set, in many cases by leading British philosophers, and discussed by senior research students who will be the leading philosophers of the near future. The series was introduced in the following terms: 'Since most readers of *Analysis* are in fairly general agreement about the aims and methods of philosophy, it has been decided to try the experiment of setting from time to time a limited and precise problem to be "solved" in not more than 600 words. By this means, it is hoped, some small, but quite definite results may be obtained. Nor should these be despised as trivial. Though small in compass, they may be large in implication.'[1]

Here are a few of the problems, selected arbitrarily:

No. 1. 'What sort of "if" is the "if" in "I can, *if* I choose"?'
No. 3. 'Does the logical truth $(\exists x)(Fx \vee \sim Fx)$ entail that at least one individual exists?'
No. 6. 'How can one wish to have been Napoleon?'

[1] *Analysis*, Vol. 12, No. 3, 1952 (Oxford, Blackwell). To relieve possible suspense, it should be added that 'solutions' of the problems will be found in succeeding issues of *Analysis*.

No. 10. 'It is impossible to be told anyone's name. For if I am told "that man's name is 'Smith'", his name is mentioned, not used, and I hear the name of his name but not his name.'

No. 12. '"All swans are white or black." Does this refer to possible swans on canals on Mars?'

No. 13. 'Is "Paradise Lost" a general name, proper name, or what?'

Now all the wealth of the Indies would not lure me into a condemnation of this activity. I have no doubt whatever that if we could solve these problems completely we should know as much as if we understood Tennyson's flower in the crannied wall.[1] But what I do regret is that, in their zeal to reach this ultimate illumination, the acutest minds of our generation should fail altogether to perceive the perplexity into which our logically amateurish scientists are plunged over problems which they could solve so simply and which have implications for the immediate welfare of mankind as large as those which their own preoccupations have for eternity. I even venture to express the opinion that if our philosophers would only take the risk of assuming—tentatively, and without prejudice—that there is (or is not) a way of wishing to have been Napoleon, and give a passing attention to these superficial problems which may nevertheless involve the possibility of the future existence of man on this planet, they would make a contribution to our common welfare more closely approximating to the responsibility which their intellectual endowments lay on them than anything they have yet done. But, I repeat, that is only an opinion.

3

And now, at long last, back to sanity again, and the question of the ether.

Let me begin with a comment on motion 'in' or 'through' the ether. This point has received much attention, and various views have been held. If light is a wave in the ether, then clearly ether must exist inside transparent bodies at least, and if it is a medium for electromagnetic influences also, it must exist inside all bodies. We must therefore ask: if a body moves, how does its relation to the ether in and around it change, if at all?

[1] See p. 179.

Fresnel in 1818 was the first to give a well-supported answer. He supposed that the ether in a transparent body was denser than that outside. When the body moved it carried with it all its *excess* of ether, leaving in its former location ether of the density normal for empty space. This was satisfactory for optical phenomena, with which alone he was concerned.

In Maxwell's electromagnetic theory the functions of the ether were extended, and Fresnel's conception met with difficulties. Hertz tried to remove these by supposing the moving body to carry *all* its ether with it. This was only partly successful, and was soon abandoned in favour of Lorentz's theory, in which the moving body carried *no* ether with it. The ether was an infinite, omnipresent, quiescent sea, and although the particles of matter could, in a sense, be regarded as made of the substance of ether, they were interpenetrated by this universal sea as though they were not there.

That seems to be the only kind of ether that meets experimental facts if motion with respect to it (whether 'in' or 'through') is to be meaningful. But—with full recognition of the fact that if asymmetrical ageing should be proved we should have to admit motion with respect to ether and so revert to Lorentz—we are both prepared at present to ignore this possibility and accept an ether, if at all, only as something to which no unique state of motion can be assigned. What can such an ether be? Your answer is: 'energy'. I take it for granted that you do not wish to reinstate a discarded conception by merely changing its name, so that you mean by 'energy' what physicists in general mean by it. We must therefore ask: what do physicists mean by energy?

Strangely enough, no one has ever been able to say: like 'life', it eludes a concise definition. Text-books tell us that energy is 'the power of doing work'. But all physicists hold that if an isolated system of bodies came to a common temperature, its energy would be undiminished but it would have lost all power of doing work. We can define energy only by saying that if we make a set of measurements on a possibly changing system, a certain combination of them does not change: it measures the *energy* of the system. That is a discovery—one of the greatest ever made in science—but all attempts to reify 'energy' (as we inevitably tend to do when we give it a name) lead to contradictions.

Let me try to illustrate this by the simplest possible case. Suppose a body, whose weight as given by a spring balance is W, is held at a height H above the ground. Now release it. It falls with increasing speed to the ground, which it reaches with velocity V, say. It is then found that (disturbing factors having been, as usual, allowed for), if M is its mass, the quantity $\frac{1}{2}MV^2$ is equal to the product WH. This is true no matter how we vary the conditions. Any solid body will do, the weight and height can be changed *ad lib.*, the experiment can be made anywhere on the Earth. Furthermore, if at any stage during the fall the height is h and the speed v, then the sum of the quantities Wh and $\frac{1}{2}Mv^2$ is equal to WH, i.e. to $\frac{1}{2}MV^2$.

We express these facts by saying that the body possesses something called *energy*, which remains constant in amount, but can change its form from *potential* energy, measured by Wh, into *kinetic* energy, measured by $\frac{1}{2}Mv^2$. At the beginning of the process, v is nothing, so all the energy is potential and its amount is WH. At the end of the process, h is nothing, so all the energy is kinetic and its amount is $\frac{1}{2}MV^2$, the same as before. The process consists of the continuous transformation of potential energy into kinetic energy.

This is perfectly legitimate and very useful, provided that we do not suppose that 'energy' is actually some physical existence which we have discovered. But that is what we automatically tend to do when we invent a noun. The initial 'potential energy' of the body is merely the man-made product of the quantities W and H which we read on the man-made and man-designed instruments which we voluntarily introduce into the situation. It is no more something which the body secretes than time is something which flows past us—a fallacy which you have very clearly exposed. To see this, suppose the experiment occurs when the Sun is directly overhead. Then, if we take the Copernican viewpoint and refer our measurements to the Sun, the body *gains* potential energy when it falls; as it nears the ground it recedes from the Sun. Is it, then, *really* gaining or losing potential energy? The question is meaningless, because there is no such 'thing' as energy: it is a term affording a convenient expression of relations between the results of certain operations of measurement which we have been clever or fortunate enough to select from the infinite number of those that are possible to us.

So when we speak of potential energy—and the same applies to all forms of energy—although it may be convenient in limited contexts to reify the concept, we must always remember that this is merely a convenience, and not transcend the limitations. It must always be possible, when we use the word, to substitute the corresponding combination of measurements: otherwise we shall be the fools of our nomenclature. Such a substitution is always possible in true physics. When the physicist says that a body loses heat energy, or speaks of the energy density of an electromagnetic field, and so on, he can in each case state the measurements necessary, and the way of combining their results, to give precise meaning to his statement. If he could not he would become a visionary.

When, therefore, I consider your proposal that space is filled with energy, quiescent or active, I see no objection to these concepts if they can be interpreted in terms of the measurements for which they stand. At present I cannot so interpret them, but I do not say that it is impossible. When I ask what purpose they serve and why we should feel prompted to postulate them, the only answer I can yet give is that they afford a mental picture of a process taking place when heat and light travel from one body to another. That is a useful function, but is it any different from the picture which the nineteenth-century physicists had of an 'elastic solid' ether of which the particles, normally quiescent, were stirred into activity as they participated in the wave motion passing over them, and then relapsed into quiescence again? That conception broke down precisely because, when it was converted from a mental image into an expression of relations between measurements, it led to contradictions. What is there in your conception that would avoid those contradictions?

In particular, do you find it possible to picture this kind of ether without introducing the idea of an identifiable standard of rest with respect to which the motions of bodies can be measured? I am able to do so only in terms of Faraday's 'rays', of which I spoke earlier (p. 65). From every atom there proceed rays in all directions, extending to infinity unless they end on another atom. If the body moves they move instantaneously with it. Every atom (the rays are an essential part of it) fills all space. Normally the rays are quiescent and give no evidence of their existence, but when the atom influences another—by transmission of light or other electromagnetic quality—they become active as a wave passes along them with the invariable velocity c.

That might indeed afford an interpretation of your ideas that would lend itself to mathematical development. I see no more *intrinsic* difficulty in giving mathematical expression to this than that residing in the problem which Maxwell solved of so expressing his version of Faraday's ideas in terms of a single, featureless, stagnant ether. Each atom would have its own 'ether'. This would not only eliminate 'action at a distance' and preserve all the picturable advantages of the ether idea without requiring a universal standard of rest, but might also relieve physics of a very unsatisfactory situation. The present electromagnetic field equations can be derived only by assuming an ether in which no one now believes. We have discarded the necessary basis of the equations, but retain the equations, floating on nothing, so to speak.

A second Maxwell, who could do for Faraday's idea of ray-vibrations what the first did for his lines of force interpreted as strains in a unique ether, might relieve us of many difficulties and give to your conceptions a justification which I am sure physicists must feel that in their present form they lack.

Though this is the only form in which I can see your conceptions appealing to physicists, I am far from saying that it is the only possible one. But theoretical possibilities inevitably make less appeal than a concrete example, and this, I think, does show that the picture you present is potentially susceptible of legitimate mathematical expression. But there are two points on which, if I have rightly understood your attitude, I think I ought to express my dissent.

You speak of action at a distance as 'wholly unacceptable' and worthy of 'scorn'. That seems to me to place a limitation on physical research for which there is no warrant. We have to give a rational account of things as we observe them, and if we can do so in terms of action at a distance while all attempts to give a rational description in terms of an intervening medium fail (I do not say that this is so) I see no reason why we should not. As you have rightly said, space is a 'fictional abstraction', and the fiction writer is master of his medium; he is as free to make it empty as full. Newton, it is true, scorned the idea of action at a distance, but then Newton did not look on space as a fictional abstraction. To him it was 'the sensorium of God', the most absolute thing in the universe, and he points out, in the General Scholium to the *Principia*, that 'bodies find no resistance from the omnipresence of God'. We do not hold such ideas today, so I do not think we can properly claim Newton's support for the idea that action at a distance is inadmissible.

Secondly, I do not think we ought to deny ourselves the use of a concept because we cannot form a mental picture of it. What we must insist on is that our concepts shall not be permanently logically irreconcilable (we may tentatively adopt incompatible ideas in different fields of study in order to make progress, but only as 'something to be surpassed'). Hence, to take an example from a case already discussed, I would not object to the statement that a space-traveller, through his motion, ages by a day while an Earth-dweller ages by 100 years, although I cannot imagine such a process; I can imagine only my astonishment if it should happen. But I do object to the statement that that can be deduced from a theory that allows the Earth-dweller, and not the traveller, to have made the movement. That is a logical contradiction. I would therefore be prepared to believe that when we have found a relation to hold good between our measurements, and have expressed this by a mathematical equation, it may be impossible to go further and say: 'this symbol corresponds to such and such a picturable state, and that symbol to that'. We may have to be content to say: 'this symbol stands for the quantity which I get when I perform a certain operation', and leave it at that. I do not think we have a right to demand what I think you mean by 'a realist cosmogony'; nature may transcend our imagination. But if it violates our reason, science and philosophy are vain pursuits.

LORD SAMUEL:

1

It is evident that we are still too far apart on fundamentals to be able to come to an agreement about ether. But it may be also my fault, because I had not made it clear in advance why my conception would not, in my opinion, be open to the objections that you raise in your pages. I certainly do regard an ether as being something physical—a real element in the real universe, and not merely a system of measurements. We may give it a place in the series beginning with material objects and proceeding through molecules, atoms, particles, pulses and lines of force, to active energy, and then to quiescent energy. But quiescent energy does not attach to matter, or be possessed or conserved by matter, or related to it in any way at all—except only that it underlies it; and when it is activated it may itself become matter, by that series reversed.

Nor can I claim that quiescent energy can be expressed in any kind of mathematical symbolism, or by any form of definition or description, because, for the very reason that it is quiescent, it presents no phenomena. At the same time, we can claim to know that it exists, because its essential nature is that it is capable of being activated, and when active it becomes the matrix from which originate all the phenomena that we know. All are patterns of active energy. I can only ask you to consider the problem yet again, but to include as the essence of it, the concept of states and the concept of patterns.

2

In my original sketch of the possible mechanism of a two-state ether, I was discussing the embargo imposed by physicists, during the last fifty years, upon any discussion of an ether of any kind because of the null result of the Michelson-Morley experiment; and I was putting forward the view that that result might have been foreseen because it was based on an assumption which was a false assumption. This was that the Earth was one thing and the ether was another, and that what was being sought was a measure of the velocity of the Earth's motion *in* and *through* the ether. I submitted that any true ether must be a continuum or plenum, existing, if at all, everywhere and always. Consequently it could not take part in itself in any movement, because there was nowhere else that it could go where ether would not be already present; and nothing could move *through* it, because movement *through* involves starting from outside and ending outside, and there was no 'outside'. In all this I was speaking only of quiescent ether, but I may not have made that clear. Wherever and whenever the ether has been activated (how that could be brought about being a separate question) those considerations do not apply. The words *in* and *through* may come into use, with respect to the activated ether in any of its patterns. This is obvious in the universe of our experience, which consists of activated ether in its patterns of sub-atomic particles, combined into atoms, and then constituting molecules and material objects.

You say (p. 87) that the conception of quiescent ether might lend itself to the locating of particular points as standards of rest which could be the basis for frames of reference for any kind of motion, thereby violating the principle accepted in relativity theory, that there can be no such thing as absolute motion. I submit that that is not so. Quiescent energy, presenting no phenomena, can supply no point of relation for any frame of reference. The moment it is activated it certainly can do so, but from that moment it is no longer quiescent energy. The words 'in' and 'through', inapplicable to quiescent ether, would return as appropriate to active ether.

3

I would borrow the word *substance* from the old philosophies, for it seems useful and etymologically apt in this conception, and I would say that an ether of activated energy is the substance underlying the whole cosmos—and its only substance. In it we should find again the universe of our perceptions, of our experience. And the theatre of study of our sciences, theoretical and practical, would reappear. The whole majestic structure of modern science, including its mathematics, would continue for the most part little affected. But it would be simplified in some important particulars, and be provided once more with a firm and self-consistent physical foundation.

You ask why I should wish to trouble with ether and other controversial propositions; what use are they likely to be in any event; I have an answer to that challenge, and a very definite one. I should like to give it here and now, but it would be far better for it to come in the concluding chapter, when it could cover the subjects still remaining for the next two sections.

4

But before leaving the ether, I would point out that you have misconceived my proposition for an alternative to the present orthodox theory of modern physics on the transmission of electromagnetic radiation and other forms of energy. That theory alleges that energy is something—its nature undefined—which is of a predetermined quantity in the universe; it may take one form or another, but there is a natural law of 'conservation of energy', which requires that, through all transmutations, its mathematical quantity can be neither increased nor diminished. The hypothesis for which I have argued, elsewhere and here, supports the acceptance of energy as a fundamental factor in the physical universe: but it does not find any ground for believing that the 'dose' of energy engaged in any particular phenomenon remains constant in quantity—that it may disappear out of our reach but it cannot disappear out of the universe: we must postulate its continuance as 'potential energy', and that 'potential energy' may 'become dynamic and then kinetic', and in that form may 'do work'.

Our alternative does not use that vocabulary at all. It does not accept that there is any such law of conservation. There may be transmutations in which the amount of energy can be measured (e.g. friction into heat), and the measurement will show the same quantity in the latter form as it had shown in the former: but we hold that in other phenomena, the energy process may be stopped, or may be exhausted and fade away, and no further continuance be observed.

In your comment you refuse to accept this alternative hypothesis. You do not say that it is wrong. You do not seek to refute it. The only ground that you give is that it does not conform to the present orthodox views of physicists. You use their vocabulary—a material object 'acquiring energy'; and then 'preserving' it as 'potential energy'; afterwards 'expending', or 'losing', or 'transferring' it. When referring to the mechanism suggested as possible for the alternative hypothesis—a universal continuum of energy, which exists in either of two states, quiescent and active, and in various patterns—you truly say that the other vocabulary, which belongs to the orthodox hypothesis, will not fit this one. Of course it will not, because this theory does not admit any of those processes or situations—a material object possessing or conserving energy, or expending, losing or transferring it.

Before passing on, however, leaving behind us a definite disagreement on one of the principal subjects of our discussion, I would make one more effort to establish my case for a two-state ether.

5

You take as an illustration a material object moving under the influence of the earth's gravity. I will do the same, and would offer as my test example one of the hydro-electric dams that are being built all over the world across rivers, great or small; they create, up above, artificial lakes as reservoirs to serve the turbines of electric power stations down below. I would ask you to consider afresh, on their merits, the theories that are offered as alternatives.

The theory as to the energy engaged, that is now taken as established, would find the beginning of the process in the heat energy of the Sun's rays. This 'sucks up' the surface layer of water of the sea or of lakes, ponds or pools, and converts it into water vapour, or clouds or mist. In the process the energy involved is transformed into an equal quantity of energy, which is then said to be 'possessed' by the H_2O molecules which constitute the clouds and the like. This remains with those molecules as 'potential energy', and follows their fortunes as they pass from water vapour into clouds, or mist, snow, or rain, and then as drops of water into rills or streams, or the upper waters of a river. So the energy, originally in the Sun's rays, is now 'possessed by' the water in the reservoir above the dam. There it may be 'conserved', for a longer or shorter time, until it is 'released'. The water is then allowed to rush down a channel to the turbines, where the energy is transferred to the dynamos in the power-station. Thereby it has been first converted from 'potential' to 'kinetic' and 'dynamic' and then 'spent' or 'lost'. Afterwards the river flows on placidly; but is able to do no more 'work', than perhaps to turn the water-wheels of small flour-mills here and there.

Now consider the alternative theory of the realist. This would start in the same way, with the Sun's rays heating water surfaces. But it would not speak of 'suction', for it in no way resembles mechanical suction. What happens is that the molecules of the heated H_2O expand, and, becoming vapour and lighter than the air immediately superposed, rise above it, the water vapour molecules mixed with it being carried up with it. The cooler air above, responding to the pull of the Earth's gravity, presses downwards and quickly changes places with the warmer air that has risen. It is not a case of the heat sucking the warm air upwards; it is the Earth's gravity pulling the cold air downwards. The effect is the same, but the force that is held to cause it is different.

The next stage of the history is also the same in both theories

—the conversion of the water molecules from patterns of vapour or dew into patterns of soft snow crystals, or hard ice as hail, or raindrops; and then into the flowing liquid water of springs, rills, rivulets, and the upper waters of the river, until it reaches the reservoir. From then on the two theories differ completely. For the realist, the whole story of the energy from the Sun's rays having been 'conserved' during migrations through vapour to clouds and thence to the reservoir where it is 'possessed' by the water; there to be converted from 'potential' to 'dynamic', and finally to 'kinetic', before being 'released' as the water rushes down to the turbines in the valley, there to be 'spent' and 'lost'—to the critic all this is sheer myth. Nothing of the sort happens as actual events in the real universe. Water below the power-station is just the same as the water in the reservoir. You may take up on a mountain walk an empty 12 oz. medicine bottle, fill it from the reservoir, and bring it down in your rucksack: you will find it unchanged; perhaps a little different in flavour, but essentially the same. From the beginning to the end of the story, the water has been altogether passive. The 'It' is incapable of *doing* anything at all. It cannot conserve or possess anything, or release, spend, lose or transfer anything. The only part that the water plays in the whole matter is its passive response to the pull of gravitation. Whatever is done is done *to* it, not *by* it. The active agent is first the Sun's heat, and afterwards, throughout, the Earth's gravity.

6

While we have been engaged in debating these theories, a very tragic event has happened in the French Alps, which gives a painful actuality to such questions as these. On December 2nd, 1959, occurred the disaster of Fréjus. A dam, which had been recently built across a little river, gave way, and the whole contents of the reservoir above, at their maximum owing to an unusually rainy season, were suddenly hurled upon the valley below, devastating the town and killing nearly four hundred people.

What had caused the disaster has been the subject of an inquiry by a committee of civil engineers of international repute. Was it due to a fault in the design for the dam, or to a failure to open the sluices in time, or the rock foundation being insecure and slipping? All that was a matter for technicians.[1] The prime cause was plain enough—the water in the reservoir had rushed through the broken dam *under the pull of the Earth's gravity*. The engineers would have no occasion to consider how the water came to be where it was; whether it came there originally through the suction of the Sun's heat rays lifting it to cloud levels, carrying energy with it, which is then taken on in the snow or the rain and becomes possessed by the water in the reservoir, and so forth. The engineers, for their special purpose, would have no need to trouble their heads with all that. They would be concerned with a single fact—that a volume of some millions of tons of water suddenly struck the valley, and that the force that hurled it was obviously the Earth's gravity.

Coming now to the essentials of our problem, we find, as so often happens, that while we are supposing we are dealing with one familiar problem, we are in fact dealing with two problems —sometimes more—tangled together, and our first business must be to disentangle them. In this case there are two phenomena, one following upon the other. The first question is, seeing that water, when it is free to move, will move downhill and not uphill, how is it that all that volume is situated at the top of the valley and not at the bottom? The problem, as I have stated it, began with the water in the reservoir as a given situation, and that is all that the engineers inquiring into the cause of the disaster need concern themselves with for the purpose they have in hand. But the physicist must not ignore the other, and prior, question. That presents a phenomenon separate from that of the burst dam and the downrush of the whole mass of water; and while the Earth's gravity has here also a leading place, some other force, counteracting gravity, must come in: and that is the influence of the Sun's heat-rays in causing the water of the surface layer of tropical or sub-tropical seas to expand. The effect is that there are fewer molecules of H_2O in a given volume of the warmer air, which consequently offers a less response to the pull of gravity. We say that the layer of cold air above is relatively 'heavier'; it quickly filters through to a lower place of rest, while the warmer air rises until, in lower temperatures, it condenses

---

[1] The inquiry has found that it was not a fault in the design or in the construction of the dam that was responsible, nor any mistake in the functioning of the floodgates. There was a fault in the rock foundation underneath the dam, and it was that that was to blame. (See *The Times*, March 28th, 1960.)

into various forms and so reaches the reservoir as liquid water. But the forces engaged are only two—the radiation of the heat waves from the incandescent Sun, and the gravity pull of the Earth. The whole of the physicists' vocabulary of the supposed peregrinations of the doses of energy—the water-vapour 'acquiring' energy, 'conserving', 'transferring', transforming it into 'potential energy', then 'kinetic', next 'dynamic', and finally releasing it to be transformed into electric current in the power-station in the valley—all this, it seems to me, is pure myth. No such processes are actually happening. It is all mere scholasticism—words, given as names for processes, being taken as conferring reality. The vocabulary is redolent of the mediaeval monasteries before Galileo and Francis Bacon. It is at least 300 years out of date. It should be recognized to be misconceived, and the sooner it is eliminated from the physics textbooks of our schools and colleges, the better.

7

Now the point that I would put to you is simple, but the answer to it may be decisive. Do you consider that what may be called the engineers' account of the event is right, or the physicists' account? If you agree that the key to the whole situation is the Earth's gravity, and not *the possession of energy* by the water, then your objection to my ether hypothesis is invalidated, your *a priori* veto is removed, and the way is cleared for an examination of the hypothesis on its merits.

When I ask what is the reason for inventing all the artificial complications required by the alternative hypothesis, the physicists would answer that it is essential in order to comply with a natural law of the conservation of energy. And if we go on to ask why we should believe that any such law exists, we shall get the answer that it is because the law is exemplified in phenomena such as we see operating in the transfer, through a number of transmutations of energy from the Sun's heat to the hydro-electric power station. Again we are offered a circular argument, which leaves us at the end where we started at the beginning.

PROFESSOR DINGLE:

I am sorry that I did not appreciate the fact that you were departing so radically from current physical ideas as your new statement makes clear. I was misled by your adoption of the word 'energy', which for the last hundred years has had a precise meaning in physics as something which, by definition, is conserved. Its various forms were thought at first to be quite independent.

What we now call kinetic energy used to be called *vis viva*. Other forms were referred to in the first half of the nineteenth century as 'forces' of various kinds, and it was not until the middle of that century, when the fact gradually came to light that when one of these 'forces' vanished another appeared in proportionate amount, that the idea arose that there was some general entity underlying all such changes, whose *substance* remained the same eternally while only its *form* changed. The name *energy*, apparently first appropriated for this entity by Lord Kelvin, came into use about 100 years ago. The word as now used is thus inseparably associated with the idea of conservation. Conservation of energy is, nevertheless, not a truism, for the separate forms of energy are recognizable and measurable independently, and it is therefore conceivable that any of them may increase or diminish in amount without any compensating change occurring in any other form. But the outstandingly important fact is that this has not been known to happen, and physicists have therefore acquired such a trust in the conservation of energy that when energy *seems* to disappear they almost instinctively take this as evidence of the existence of energy in a new form whose independent manifestation further research may be expected to discover.

A recent example of this is the discovery of the very light, uncharged particle known as the *neutrino*. In certain experiments there appeared to be a lack of balance between the total amounts of energy at the beginning and at the end. The balance would be restored if unobserved particles having certain properties were involved in the experiments, and these properties were in fact such that observation of the particles would be exceedingly difficult. Physicists therefore had considerable confidence in their existence, and the later confirmation of this by independent observation caused no surprise and was regarded as little more than a formality. If this seems to you to show overconfidence in generalizations which should always be regarded as tentative, I should be inclined to agree; but still, it does indicate that there are very strong grounds for associating the idea of conservation with that of energy, and certainly if conservation were ever found not to be universal, the idea of energy would lose much of its significance and might ultimately be abandoned altogether.

So I think that if the concepts which you have introduced under the names 'active energy' and 'quiescent energy' take no part in any conservation law, either of them being able to appear and disappear without any equivalent change in the other, it would be better if you gave them different names, for physicists would automatically tend to assume that, being forms of energy, they would be subject to the conservation law. However, now that you have explained the position, we need not for our present purpose invent new terms, so I will try to reconsider your suggestion without confusing it with incompatible concepts belonging to a different system of thought.

But when I do that I immediately begin to ask myself: what is gained by introducing these terms? What do they do to enable me to see in a situation something that is not immediately obvious in the mere existence of the situation itself? You have cited the case of the Fréjus dam. Before the accident, all that anyone could observe (reducing the matter to its simplest essentials) was a quantity of water behind a material structure, on the other side of which was a town at a lower level. Suppose someone quite ignorant of science to be placed in front of this situation and asked to say what he could about it. Then the statement I have just made would exhaust his possibilities: he would have no conception of what might follow or of how to attempt to determine it. But through modern science it would be possible (in practice, of course, it would be very difficult in this particular case to make the necessary observations with sufficient precision, but that is not relevant to the present point) to make such measurements as would enable one to say that when the water had accumulated to a certain amount the dam would yield; that then a precisely specifiable amount of damage would be done to the town below; and that eventually the water would proceed in a certain direction with a certain speed and with a certain increase of temperature. Furthermore, it would be possible to know what to do to prevent all this from happening. And this would be possible because we could know the potential energy of the water, the energy necessary to break the dam, the kinetic energy the water would acquire at the lower level, and the equivalent amount of heat energy: in short, because of our knowledge of the conservation of energy.

You will, of course, not question this, but you will say that it does not *explain* the occurrence. As an explanation you call it 'a circular argument, which leaves us at the end where we started at the beginning'. But it is not a circular argument. At the beginning we were in the position of our unsophisticated visitor who could simply describe the initial situation. Because of the physicist's law of conservation of energy, we could have predicted the *final* situation and, if the necessary observations had been made, prevented it. There is nothing circular about that. Apart altogether from the humanly beneficial aspects of this knowledge, we have the knowledge itself, the fact that the original situation as we can observe it contains the potentiality of the final situation, and we know in what circumstances that potentiality can become actuality. That is knowledge, not only in the modern scientific but also in the old Aristotelian sense of the term, and it would not have been possible without the physical law.

I confess that I cannot see what the account of the matter in terms of quiescent and active energy can offer us that is comparable in value with this. I do not see that it enables us to predict anything. When something happens it gives an account of that something in its own terms, but we have to wait for the happening before the account becomes possible, and I do not see what then it adds to our understanding. We are back at our fundamental difference regarding the universe as it is in itself and as it is for us. You want to learn about the real universe that lies behind the appearances, and your quiescent and active energy belong there; they are the cause of what we observe. But what physics does is to look *into* the appearances, not behind them. It sees what potentialities are contained within them, and leaves the question of what lies behind—the universe as it is in itself—as something which, even if its existence be granted, is not relevant to its activities.

This question was very much to the fore in the seventeenth century, and almost in the same connection as that in which we are now discussing it. The treatment of gravitation by Galileo and by Newton illustrates the point excellently. In Galileo's *Dialogue on the Two Great Systems of the World*, the question is raised, 'what moves the parts of the Earth downwards?' 'The cause of this,' says Simplicius, the Aristotelian, 'is most manifest, and everyone knows that it is gravity.' 'You are out, Simplicius,' says Salviatus; 'you should say that everyone knows that it is *called* gravity, and I do not question you about the name but about the essence of the thing. Of this you know not a tittle more than you know the essence of the mover of the stars in gyration.' Here the contrast between the phenomenon—the falling of bodies to the ground—and the *cause* of the phenomenon is brought out clearly. We do not know the cause of the phenomenon, and merely to give it a name does not relieve our ignorance in the least.

In the next generation the problem was taken up by Newton. He realized clearly the difference between the phenomenon and its cause, and he did not imagine that he had discovered the cause when he had called it by a name. But, on the other hand, he did not make it his business to look for the cause—that, he said, 'is what I do not pretend to know': he confined his attention to the phenomenon. 'To tell us,' he says, 'that every Species of Things is endow'd with an occult specifick Quality by which it acts and produces manifest Effects, is to tell us nothing: But to derive two or three general Principles of Motion from Phaenomena, and afterwards to tell us how the Properties and Actions of all corporeal Things follow from those manifest Principles, would be a very great step in Philosophy, though the Causes of those Principles were not yet discover'd.'

The 'cause' of gravity is still not discovered, but we realize now that the only kind of 'cause' of gravity that we could conceivably discover would be a relation between it and some other phenomenon such as electromagnetism. We might then be able to say that the cause of gravitation is the resultant electromagnetic field of sub-atomic particles, or something of that kind, and that would connect the fall of an apple to the ground with, say, the aurora borealis, just as Newton connected it with the orbital motion of the Moon; i.e. it would relate apparently independent phenomena together, but it would not get *behind* any of them. It would describe the universe as it is for us; it would not penetrate to the universe as it is in itself.

To come, then, at last to your specific question—do I consider that the engineer's account (that the cause of the disaster was the Earth's gravity) or the physicist's account (that the cause of the disaster was the possession of energy by the water) is right?

—I would say that, as so expressed, neither account is sufficiently complete to throw much light on the matter, but that, when properly amplified, both accounts may be right because they afford alternative but equivalent descriptions of the situation of the only kind that can add significantly to our knowledge. Forces, like gravity, and energy are not mutually exclusive alternatives between which we must choose, as they would be if they were objective existences in an objective universe. They are concepts which we define, and we have defined them in such a way that the same phenomena can be described in terms of interaction of forces or the transformations of energy. For instance, it is possible to 'explain' why the surface of a liquid resting in a vessel must be horizontal either by considering how the force of gravity acts on the various parts of it, or by using the principle that the liquid will come to equilibrium in the condition in which its potential energy is a minimum. As picturable processes these are quite distinct accounts: in one we make no appeal to the concept of energy and in the other no appeal to the concept of force. But to ask which is 'right' would be like asking whether these words are printed in a book or in *un livre*.

So I think that by divorcing your conception of quiescent and active energy from energy as physicists now conceive it, you are robbing it of a possibility of usefulness which it might otherwise possess. As I have mentioned earlier, it seems to me quite possible that an 'ether' consisting of the interpenetrating 'lines of force' that Faraday envisaged might very well become a useful physical conception, and the difference between the unexcited lines of force and those along which vibrations are travelling might be conveniently expressed in terms of quiescence and activity. In that case your conception might well become a part—indeed, a fundamental part—of physical theory. It would seem to me most undesirable to annul that possibility by seeking for some metaphysical reality to which it would be impossible for physicists to give assent.

LORD SAMUEL:

1

I am grateful for your criticism of the ether hypothesis that I have been putting forward. With perfect clarity you have done what no other scientist, so far as I am aware, has yet taken the trouble to do, and have presented a reasoned statement of objections. But I still cannot agree.

I think that the illustration of the burst dam at Fréjus brings us to close quarters with our problem, and I will take up the argument about that where you leave it. And here we must part company at the very outset.

You say (p. 98): 'Before the accident, all that anyone could observe (reducing the matter to its simplest essentials) was a quantity of water behind a material structure, on the other side of which was a town at a lower level. Suppose someone quite ignorant of science to be placed in front of this situation and asked to say what he could about it. Then the statement I have just made would exhaust his possibilities.' I venture to suggest that is not a correct statement of the initial situation. It does not exhaust the possibilities that would lie before a non-scientist observer—or a scientist: for it leaves out altogether a factor that is of fundamental importance. It leaves out the Earth's gravity. This element is so obvious, so universal, we are so accustomed to it as part of our experience at every moment of our lives, that we may easily come to take it for granted as an ever-present background and to forget it.

I would ask you to amend your description by wording it thus: 'Before the accident all that anyone could observe, reducing the matter to its simplest essentials, was a quantity of water behind a material structure, on the other side of which was a town at a lower level—*that water being continuously subject, like every other material object on the Earth's surface, to the pull of gravity.* That would exhaust his possibilities.'

Before the dam was built the little river used to flow down into the valley: it did so because water, if unimpeded, will always flow downhill under the pull of gravity. The dam was built; the flow was held up and the water accumulated in the reservoir that had been formed, with its rocky floor encircled within the upper slopes of the valley, and completed by the construction of the dam. That had to be strong enough to withstand the weight of so many thousands of tons of water pressing against it. The question arises—what do you mean by 'weight'? And the answer is — weight is the response of any material object, lying on or near the Earth's surface, to the pull of gravitation.

I think you will agree so far, and will not dispute that the

Earth's gravitational pull is a physical reality. That being so, I submit that we ought to go further and to realize that in the Fréjus situation the Earth's gravity is the dominating factor.

You do indeed mention gravitation, but much later on (p. 99), and only incidentally, in order to bring in the fact that neither Galileo, nor Newton, nor our present-day physicists, have been able to discover and explain what gravity really is. The phenomena which it produces are part of human experience, but its own mechanism, the causes of those phenomena, we do not know.

To give it a name, 'gravity', would not be enough if we stopped there: it would not carry us any further. It would leave us, to quote your useful phrase, 'the fools of our nomenclature'. Meanwhile, you say, we must do our best to establish relationships, particularly by way of mathematical measurements, between gravity and other physical facts within our experience.

I agree with respect to the futility of giving names and thinking that you are thereby extending knowledge. But I do not consider that the absence of a knowledge of causes is of much importance for the purposes of the present discussion. I should myself say that the essence of the matter lies in the fact that the Earth itself is a magnet: but that also would do no more than generalize the problem, and leaves it re-stated as 'What is magnetism?' When we ask what was the force that hit Fréjus on December 2nd, 1959, my substantial answer would be 'the Earth's gravity', and that it does not matter at all whether Newton, or anyone else, had been able, or had not been able, to explain what gravitation is in itself. Water flowed downhill, and not uphill, indifferent to mathematics, and long before mathematicians, or any human beings at all, had appeared on this planet to philosophize about it.

As you emphasize, however, mathematics and measurements are indispensable when we come to the practical measures needed to deal with the situation. Engineers had to calculate what was the weight of the volume of water which would press against the dam, and what would have to be the strength of the wall needed to resist it.

So far I agree, but I would claim that this argument supports what I call the engineers' theory rather than the theory that it was the water that was responsible through *possessing* energy and transferring it.

2

Engineers are able to make the measurements that are needed without bringing in those ideas at all. They know what are the cubic dimensions of their reservoir; they know from their gauges what is the volume of water at any significant time; and they know — one gallon of water weighing 10 lb. — how many thousands or perhaps millions of tons weight would be involved. All that is an elementary part of their business. But they have no need to know how to define the term 'gravity', and no reason to attempt to make any calculation how much potential energy was possessed by that amount of water, how much was conserved so long as the dam stood, and how much would be released and spent when the sluices were opened, or if the dam should burst.

Is it not strange that the present orthodox school of physicists should, on the one hand, be constantly claiming that they are even more rigid than any of their predecessors in asserting that science should only deal with the results of observation and experiment, subsequently verified; and, because of that firmly held creed, they should even have been ready to revolutionize their own science, to abandon any fundamental belief in causality, and to find a substitute in mathematical probabilities; and yet, on the other hand, are willing to acquiesce in the pupils in our schools and colleges being taught as true this fantastic figment of the energy in the Sun's rays being transferred to clouds and thence to snow or rain, then to water in a mountain reservoir, and finally to reappear as electricity in a hydraulic power station? How do observation, experiment, measurement and verification come in there? Can anyone imagine any institution for scientific research undertaking an inquiry in order to trace the wanderings of the Sun's energy as it starts from the evaporating surface of some tropical seas, then being carried by the prevailing winds over the ocean and over the land to the upper slopes of the Alps, there to be acquired and possessed by the molecules of water in the rivulets and streams of the valley above Fréjus, and finally be released, with those tragic consequences that we know—and all this because it is necessary for the law of the conservation of energy to be obeyed, and that there is no other way in which nature can do that? If it were frankly recognized that the simple and obvious explanation, that is accepted and acted upon by the practical men who handle the situation on the spot, is true, namely, that the force engaged is the Earth's gravity and nothing else—what a relief that would be to everybody concerned!

All through our discussions you have rightly laid emphasis on the many mistakes that have been made by mankind in the course of the evolution of human knowledge, and how necessary it has been to be wary of accepting theories even if they seem to be obviously true and have been tenaciously held for hundreds or sometimes thousands of years. You have instanced, as a flagrant example, the theory held by ancient and mediaeval natural philosophers (now called scientists) that material objects 'possessed qualities' that made them what they were: a stone had solidity, hardness, weight, coldness, colour. But with the rise of modern science after Francis Bacon, this was discarded. It was realized that the word 'possessed', in that context, was meaningless, and that all those 'qualities' were really phenomena due to various natural processes. May it not be that present-day physicists are falling into exactly the same mistake with regard to energy, in their use of the word 'possessed' and the whole vocabulary proceeding from it? Are you sure that you are not defending here the last surviving relic of mediaeval Scholasticism?

3

I would pass on now to your criticism (pp. 96-101) of the hypothesis I had put forward of an ether consisting of energy, existing in either of two states, quiescent and active, and in various patterns, waves, particles, possibly 'lines of force', and maybe others as yet unconceived. That criticism may well be quite valid in itself, but my answer is that it does not relate to my hypothesis. You write (p. 98) that 'if the concepts which you have introduced under the names "active energy" and "quiescent energy" . . . either of them being able to appear and disappear without any equivalent change in the other . . . ' And again you say (p. 99), 'We are back at our fundamental difference regarding the universe as it is in itself and as it is for us. You want to learn about the real universe that lies behind the appearances, and your quiescent and active energy belong there; they are the cause of what we observe. But what physics does is to look *into* the appearances, not behind them. It sees what potentialities are contained within them, and leaves the question of what lies behind—the universe as it is in itself—as something which, even if its existence be granted, is not relevant to its activities.' I quote this in full because I think that there you put your finger on the central point. We are indeed back at our fundamental disagreement. I am trying all the time to get an understanding (mostly through 'legitimate inferences') of the cosmos as it is in itself, while you say that 'what physics does is to look *into* the appearances, not behind them'. You go so far as to say (p. 99) that the existence of the universe is not relevant to its activities, although it must be the prior cause of all of them. That is surely carrying philosophic scepticism and agnosticism beyond permissible limits.

For my own part, in the present instance of quiescent and active energy, I do not think the question is one of 'appearance' and 'disappearance' at all. Those words do not apply. I have never contemplated any natural process in which some portion of quiescent energy should disappear out of existence, and then, at once, or later, reappear into existence as active energy. If, however, you understand the situation to be disappearance, not through ceasing to exist, but through being lost to human perception, then, I would repeat, that I am not concerned, at this stage, with human perceptions at all. I regard that as a digression and a confusion of the issue, and I am not prepared to wander off along that path.

4

Let me then re-state, in the fewest possible words, what the hypothesis—if it can yet be dignified by that name—or speculation, in fact is.

The physical universe is conceived as consisting at bottom of a single element. (The expression 'at bottom' should be used with reserve. 'Beneath every deep a lower deep opens,' Emerson said, and we should not attempt to preclude further speculation and research, leading possibly to further discovery.) But, with regard to this ether, you have asked whether I conceive it to be a physical 'something'. Certainly I do; and more than that, I conceive it to be, not merely 'something', but everything—'the seat of all phenomena', as J. J. Thomson said. (We are still reserving, of course, life and mind, and Deity.)

This ether would be normally quiescent, and *in that state* producing no phenomena. It cannot therefore be described or defined. Nevertheless, we infer that it exists because its essential nature is a capacity to be activated. It may afterwards relapse into quiescence, immediately or after an interval; or it may continue active indefinitely. You say nothing as to the possibility of those two states, or as to the conception of states in general, although it is one of the most common of our experiences—the state of hot or cold; water being in a gaseous, liquid or solid state, and the like. You speak of appearance or non-appearance, existence or non-existence, as though they were the only forms of differentiation in nature. But the differentiation between states is not to be ignored.

There may be a further differentiation of active energy into patterns. All this your division into 'appearing' and 'disappearing' ignores. But it is the essence of the whole matter. If you or I go to bed at the usual hour and fall into a normal state of mental inactivity, we do not therefore disappear, to reappear in due course when we wake up. Or a bear or a bee when it hibernates; or even a railway locomotive, or any other piece of machinery, when it is started and stopped.

5

Finally I would give one more illustration in the hope—although now only a faint one—that it may help to carry conviction. It was a trivial occurrence, in actual experience and far removed from the tragic overtones of the Fréjus disaster.

The house in which I live has a rather awkward staircase between the ground floor and the upper storeys. Not long ago, a maid, carrying a tea-tray down too fast, tripped and fell; fortunately she was not injured, but the tray was flung forward, and everything on it was smashed, including some of the few surviving cups and saucers of a valued set. Now suppose that you, H.D., should have chanced to come in while I was discussing with the family whether I should try to get replacements or should buy a different set, and I should say that here was an opportunity for us to put to a simple test our differences about potential energy. As to the actual situation there would, I imagine, be no disagreement. Someone, carrying a tray, fell downstairs, and the things on the tray were broken to pieces.

That was a real event in a real universe, of the same order of reality as our own bodies and minds, and as the existence of the Earth and its surface, and of that area in it which we call London, and this house as part of it. The question I ask is, what was the cause of the tray being flung into the air when the maid tripped, and the things on it being broken into pieces? This raises a general question in practical and theoretical physics to which you and I are giving different answers.

I say, simply, the cause was the Earth's gravity. You give a more elaborate answer. You say that there is a natural law of the conservation of energy, which is of universal application. In this particular case, the tray with the things on it had become possessed of a modicum of 'potential energy of position' through someone having previously lifted it from the ground floor level to an upper storey. When it was flung forward that energy, from potential, became 'dynamic', and then 'kinetic'; then, at the moment when it struck the floor, the energy was released, and was spent in breaking up the cups and saucers.

To this I answer that I agree there are many cases in which energy can be transmitted from one pattern to another; for example from friction into heat or from hydraulic power into electricity; but that it is a wrong assumption to suppose that the principle is universal. In a case such as this it cannot be justified by observation or experiment. It cannot be related to anything else by mathematical measurement. In fact, the whole alleged process never happened, and the words purporting to describe it were mere words without significance.

I contend, further, that it is all an unnecessary complication, for the fact of the Earth's gravity is there to give a sufficient explanation. I say, therefore, that the theory is true, and that the alternative theory is false.

You say in answer . . .

But what will you say?

PROFESSOR DINGLE:

I accept your amendment of my description of the situation at Fréjus prior to the accident, provided that by 'the pull of gravity' you do not mean gravity as the scientist conceives it but merely the tendency of the water to flow downwards if it gets the chance. When you say 'like every other material object on the

Earth's surface' you are going beyond experience and introducing the Newtonian concept of *universal* gravitation, which was very different from the ideas it displaced. When Cleopatra says:

> I am fire and air; my other elements
> I give to baser life

she is claiming exemption from gravity: fire and air, unlike earth and water, were subject to levity. Their tendency was to rise—'as the sparks fly upward'. However, this is not important, I think, for our discussion: it is quite true that our science-free spectator can be allowed to know that the water exerts a pressure on the dam.

But from the very fact that the pull of gravity is acting all the time—before as well as after the accident—it is powerless to tell us anything about the accident itself. You say that the force that hit Fréjus on December 2nd, 1959, was 'the Earth's gravity'. But the Earth's gravity was there on November 2nd, 1959, yet it did not hit Fréjus. Hence something else is needed to explain its different effects on the two occasions. What I cannot see is how the conception of quiescent and active energy helps us here. Granted that energy was quiescent on November 2 and active on December 2, how does that help us to understand why the change took place? How would it have enabled us to know that the change was going to take place unless something was done about it? Until it can do that I cannot see that it does more than give names to the conditions after we have observed them.

I admit that there are some rather naïve scientists who seem to think that energy is something 'possessed' by a body, that changes form when physical events happen and can be regarded as a causative agent with respect to those events; and others, who know better, sometimes so express themselves in addressing the non-specialist; but I do not think they are now typical. Certainly I would not defend that view, and I would agree that it was a relic of the less reputable aspects of mediaeval scholasticism, though unfortunately not 'the last'. But scientists in general know that energy is merely a generic name for certain measurements, and when they say that a particular combination of measurements that could have been made before the accident is numerically equal to a combination of different measurements that could have been made after the accident, they do not mean that the energy corresponding to either of those combinations of measurements caused the accident to occur 'and that there is no other way in which nature can do that'. Nor would they object to the statement that the cause of the accident was the Earth's gravity—so far as it goes. We do not have to choose between 'energy' and 'gravity' as the cause of the accident. All that we do in metrical science is to find relations between measurements of different kinds, and the relations we have discovered would have enabled us, if certain measurements had been made on November 2, to have said that, under the conditions that in fact existed on December 2, the dam would break. Whether you say that then the Earth's gravity would overcome the resistance of the dam, or that the energy of the water would be transformed from potential to kinetic, is a matter of taste. There is no reason why you should not say both; they do not contradict one another.

The situation is exactly the same in the adventure of the tea-tray. You say the cause of the tray being flung into the air was the Earth's gravity. But the Earth's gravity was acting also on the numerous similar occasions on which I presume the maid did not trip, but the tray was not then flung into the air. So if I were asked why it was on this occasion—still keeping within the 'order of reality' which you have indicated—I should have to say, 'because the maid tripped'. Of course, one can go further and ask, 'why did the maid trip?', and then the reply would have to be, 'because in her haste she made a false step' or 'because of a rent in the carpet' or whatever it might have been. We can then ask, 'why did she hurry?', and so go on for ever—still keeping within the same universe of discourse.

But when you bring the Earth's gravity into the matter you are going outside that universe of discourse. What you are really asking then is, 'why is it that when a tray is released it moves downwards?', and that question would arise if the maid had never been born: it is quite independent of that particular event. I should not object to the answer, 'because of the Earth's gravity'. Nor should I object to the answer, 'because a mechanical system tends towards a state of minimum potential energy'. They are different ways of expressing the same general characteristic of phenomena, and both the Earth's gravity and potential energy are simply conceptions which we form to facilitate such expression; they are not possible 'realities' of which it is sensible to say that one 'exists' and the other does not. This universe of discourse is that of theoretical science. Still remaining within it, we can, as before, go further and ask, 'why do bodies gravitate towards the Earth?' or 'why does a system tend towards a state of minimum potential energy?' We cannot yet answer those questions, but, as I said before, the only conceivable answer within this universe of discourse would be in terms of some other phenomenon like electromagnetism. The answers to the questions that arise in this universe of discourse we call 'laws of nature'.

But we could now proceed to a third universe of discourse and ask, 'why are the laws of nature what they are?' That is a question which I regard as futile. This third universe of discourse is your 'universe as it is in itself'. The laws of nature describe the universe as it is for us, and they cannot take us outside that realm. The only conceivable way in which we could reach this 'universe as it is in itself' is by obtaining complete knowledge of the laws of nature, when we might find that the universe which they describe is the only possible universe that could form a consistent unified whole. That is what I meant earlier when I said that I was willing to grant that we might end by discovering 'the universe as it is in itself', but could not assume it earlier. It would be identical with 'the universe as it is for us' when we know everything completely.

So we have (1) the world of particular events; (2) the world of general laws which particular events exemplify; (3) possibly a world of pure reason from which (2) inevitably and uniquely follows. (Incidentally, Eddington thought we had already reached (3), but to maintain this he had to eliminate all experience except that of measure-numbers and to suppose that there was nothing more to be learnt even within that narrow realm of experience. I do not think we can now take that claim seriously.) Where in this framework can we locate your ether? I see no place for it but in (3), the universe as it is in itself, for you describe it as 'everything—"the seat of all phenomena"'. But laws of nature, manifested in (2)—the universe as it is for us—are to be expressed in terms of active and quiescent ether, which are different 'states' of the substance which in (3) is 'a single element'. I cannot see the necessity for (3), for if these two states are recognizably different in the universe as it is for us, it does not seem to me to add anything to our knowledge to say that they are really one at bottom. But that is an old story, which we need not repeat. My difficulty now is that, within the universe as it is for us, I cannot see how the description in terms of quiescent or active ether (or energy) does more than describe each event independently of other events. I cannot see that it establishes any relation between events. It gives names to the particular events in world (1), but does not advance into the generalizations that we look for in world (2).

Let me state my difficulty directly in terms of the succinct statement of your hypothesis which you now give. You say that active energy and quiescent energy are different states of the same basic ether, which may be sometimes in one state and sometimes in the other. (By the way, when I spoke of those forms of energy as 'appearing' and 'disappearing', I did not intend to imply their direct apprehensibility by human perception. I merely meant, in a general way, that the form of energy in question was operative or not in the situation considered, just as one might say that a man's courage disappears without implying that anything is lost to our senses. But I withdraw the words if they are likely to be misunderstood.) But the analogy you give to explain the meaning of these states is ambiguous—'the state of hot or cold'. 'Hot' and 'cold' are adjectives: we can speak of a hot poker and of a cold poker, and refer to these as different states of the same poker. But that does no more than give names to things that make different impacts on our senses. We make no advance in knowledge in this case until we introduce *nouns*— viz. 'heat' and 'cold'—and describe the various phenomena that we observe in terms of the supposed properties of these postulated entities. (Nowadays, of course, we regard 'cold' as merely the absence of 'heat', but in the eighteenth century the question of its separate existence was the subject of lively debate.) We can state how the amount of heat is measured, what happens to it when the body in which it is located undergoes various changes, in what circumstances it will become 'latent', when it will be available as a source of work and when not, and so on. All these statements describe properties of *heat*: there must be some material body or bodies in which the properties are manifested, of course, but it does not at all matter what those bodies are. The laws of thermo-dynamics, the relation that we call Clapeyron's equation, and such things do not describe the poker or any other object; they describe 'heat'. Granted that 'heat', as something 'existing' in the sense that the poker exists, is a metaphorical term, still the metaphor holds good: it is as true to say that the heat is available for driving a steam engine as that the poker is available for stirring the fire. And our statement of the laws or the properties or the behaviour of heat enables us to relate together a whole mass of phenomena that seem to have no connection with one another at all. Give a physicist a quantity of gas and, after measuring how much heat is needed to raise its temperature by a given amount under different conditions, he will tell you how fast a sound wave will travel through the gas. Who, from the character of the phenomena themselves, would have thought such a thing possible? And it has nothing whatever to do with the identity of the particular gas. The physicist need know nothing at all about that; his prediction will be right because of the properties of heat, not of matter.

Now do active energy and quiescent energy correspond to the hot poker and the cold poker or to heat and cold? If to the former then they only give different names to phenomena which are manifestly different. If to the latter, then they are the *potential* means of increasing our knowledge, but until they are specified in much more detail they do no more than say that a hot poker is a poker that displays heat. What further knowledge, for instance, does it give us about the propagation of light to say that it is a progressive alternation between quiescent and active energy? A modern physicist could agree to that, but he would describe those entities more specifically in terms of electric and magnetic intensity and so, from our knowledge of electromagnetism, enable us to say a great deal about how the wave will travel and behave in different media, what pressure it will exert, and so on.

It seems to me again that at bottom the difference between us is in what we demand of a scientific theory. To me it must be something that shows relations between phenomena that superficially appear independent of one another. You are less concerned with that than with getting behind each phenomenon to its cause in the real universe. Your active and quiescent energy appeal to you because they are real things, distinct from the physicists' conceptions which are merely names for mathematical quantities. I think you have a strong case against the non-philosophical, naïvely realistic physicist, who unconsciously accepts your real universe behind phenomena but peoples it with things like time, space, tensors and so on, and thinks that when he has given his symbol the value O, he has discovered a reality called nothing that is continually engaged in creating something. You see through the hollowness of that pretension. But I cannot admit this real universe behind phenomena. To me, just as there is no wealth but life, so there is no reality but experience, and in this case experience is represented by the numerical results of certain operations of measurement. (I do not want to imply that reason is 'unreal' in any derogatory sense. It is a sort of inner light, and is, of course, absolutely essential. Perhaps, to attempt the impossible task of avoiding all misunderstanding, I ought to say that experience is the only *objective* reality, using 'objective' in the grammatical sense and not as the equivalent of 'external'.) Time, space, mass, energy and all such terms are concepts which we employ to give verbal expression to the relations we find between our experiences. But as such, and apart from matters of technical detail, I cannot find any fault with them. I cannot see that anything is gained, and I think much is lost, by exchanging them for quiescent and active energy, because they do enable us to express relations that we could not at present express otherwise, and by suggestion point the way to the discovery of other relations. That seems to me to sum up my position in this matter.

LORD SAMUEL:

1

I have two or three minor comments on your new piece, and afterwards one major comment. The first relates to your objection that, before the flood at Fréjus could have been let loose—and similarly in all other such cases—some cause was needed other than the continuous process of gravity: but I took for granted that it would be understood that, when I spoke of the water in the reservoir being moved downwards, that implied *if its motion were not impeded*. If a hard frost were to block the outlet with ice, or if a long drought dried up the reservoir, it is obvious that no such flow would happen, and the turbines of the power station could not operate. Similarly if the sluice-gates were kept closed. I did not say that gravity was the only cause of the disaster. No event of any kind anywhere is ever the effect of a single cause; always of a combination of several; and if we go back to the causes of the immediate causes, and so on *seriatim*, we may soon get to thousands and then to millions. When writers on the philosophy of causation begin by positing 'A cause

A is followed by an effect B'—they soon find themselves in logical difficulties in trying to account for the intrusion of any kind of novelty; but they are positing something that never has happened, and never can happen. Always there is a combination of events before another event—an effect—is caused; and the causality lies in the combination.

What I said, for the purpose of my illustration, was that the Earth's gravity *was the only force* engaged when the town of Fréjus was struck and devastated, and that therefore the notion of 'potential energy of position being released and becoming kinetic' does not have to be invented in order to account for the disaster. The same applies to the normal working of the reservoir through operating the sluice-gates. The force of gravity is there all the time, whether the sluices are open or shut, or whether the dam is intact or broken; but it cannot be effective to move the water, because the volume of water is contained, and immobilized, by the sides and floor of the reservoir, unless and until there is an opening to let it out, either normally and gradually through the sluices, or abnormally in a flood through the burst dam. Perhaps I ought to have made that clear, but I hope that now it will be accepted, and that I need not surrender to the advocates of 'potential energy of position which becomes kinetic'. If someone offers you a glass of sherry it is assumed by both parties that he will first have uncorked the bottle; only then will the force of gravity be free to cause the wine to flow out of the bottle into the glass. If he has not uncorked it and has no corkscrew at hand, his hospitable offer will be ineffective, or even derisory.

2

I am very glad to note, however, that you say (p. 110) that you would not object to the answer (in such cases) 'because of the earth's gravity'. Also that you say (p. 109) 'I admit that there are some rather naïve scientists who seem to think that energy is something "possessed" by a body, that changes form when physical events happen and can be regarded as a causative agent with respect to those events . . . but I do not think they are now typical. Certainly I would not defend that view, and I would agree that it was a relic of the less reputable aspects of mediaeval scholasticism, though unfortunately not "the last".' But I would not like to leave the matter there. For this doctrine which you say is held by 'rather naïve scientists, who are not typical', is the theory which is now being taught (unless there has been some change quite recently) in the elementary and advanced science classes in the schools and colleges of this country, under the authority of the Board of Education and with the acquiescence of the leaders of theoretical and practical physics.

While I was writing about this in my little book *Essay in Physics*, I had the opportunity, thanks to the courtesy of the Education Officer of the London County Council, of seeing some forty elementary and advanced text-books that were being used in the London schools. In a note I gave some typical examples from those which dealt directly with this point. As it is very relevant to our present discussions, I repeat that note textually here. I do not recollect that any of the text-books made reference to gravitation.

' "The energy which a body possesses as the result of its motion is called Kinetic Energy. A brick on the table has more energy than a brick on the floor, though both are stationary; for if the brick is allowed to fall from the table to the floor, it can *do work* in doing so. Energy which a body possesses as a result of its position is called 'Potential energy'." (F. Sherwood Taylor, *General Science for Schools*, Part I, p. 115, William Heinemann, Ltd.) "Energy is the name given to the capacity or power of doing work, and obviously anything which can produce motion possesses energy, e.g. a wound watch-spring, when released moves the cog-wheel system of a watch. Stored-up energy of this kind is called Potential Energy." (R. G. Shackel, *Concise School Physics*, p. 77; Longmans Green & Co.) "When we wind any clock we put into it quickly a quantity of potential energy which afterwards comes out very slowly in working the clock." (E. N. da C. Andrade and Julian Huxley, *Things Around Us*, p. 91; Basil Blackwell, Oxford.) These quotations are not intended as in any way a criticism of the compilers of the text-books: they can do no other than present, in a shape suited to students, the facts, conclusions and theories that are currently accepted by the general body of teachers of the science in this and other countries. But the extracts make it evident that physics is still teaching that "potential energy of position" is an actuality and not a myth; and would lead us to believe, for example, that any rock-boulder near the top of a mountain must "possess" more energy than a similar boulder at the bottom, because, if set rolling, it could in the technical sense "do work".'

3

Another point in your latest contribution is your questioning of the analogy I used to illustrate the meaning I attached to the term 'state'. I said that to speak of the ether, or energy, being at one time in a state of activity and at another time in a state of quiescence was like speaking of a material object as at one time hot and at another time cold while itself remaining the same object. This you regarded as ambiguous because those statements 'describe properties of heat'; the physicist's laws of thermodynamics, and his mathematical measurements, 'do not describe the poker or any other object; they describe "heat" '; and you ask 'Do active energy and quiescent energy correspond to the hot poker and the cold poker or to heat and cold?' To this I would reply that I do not regard 'heat' as something extant in the universe of its own right; nor yet as a 'quality' or a 'property' appertaining to some material object. You will recall the efforts of the chemists in earlier centuries to discover an element, possibly a liquid, which they named 'caloric', and which they were convinced existed and produced the phenomena familiar to us as heat. Also the sustained endeavours of the chemists in the later years of the seventeenth century, particularly in Holland, to identify heat as a substance that they called 'phlogiston', producing evidence which was sufficient to convince many of the leading chemists of Europe.[1] They had indeed found something, which could actually be revealed by weighing; but it proved to be what they began to call a 'gas', and which was indeed a negative form of what we know so well as oxygen.

I do not suggest any attempt to renew researches of that kind. I would regard heat simply as one of nature's processes in the given universe. It is apparently at the atomic size-level, and in some form of vibration, or other kind of agitation, of atoms. Physicists are usually ready to recognize it as such, and to give it the name of 'thermal vibration'. To use the word 'thermal', does not of course confer reality. Nor would it carry us any further

---

[1] I happen to have been re-reading lately Thackeray's *Vanity Fair* and found, in his description of one of his characters that she was accustomed, whenever she felt unwell, to have recourse to 'antiphlogistic medicine'.

for those who accept an ether hypothesis to say that the heat process is an ether phenomena—for the same would be said of every material phenomenon. But that is the answer that I would give to your question. It does not seem to me ambiguous. The notion of 'states' is of the essence of this ether hypothesis. If anyone says 'what do you mean by "state"?' we may properly answer that everybody knows—you have only to put your hand on a hot pipe in a radiator to know the difference between the pipe's state of heat and its state when cold; or you may think of the difference between a person's mental state when awake and when asleep. The pipe is the same pipe, the person is the same person, and the ether is the same ether, when in one state at one time or in another state at another.

4

There are other points in your latest instalment where you give a welcome approval to some of the theses for which I have been contending, here and elsewhere. You even encourage me to hope for the possible acceptance by physicists generally of some form of ether hypothesis, with energy as its main feature, although you attach to that two conditions to which I myself would not be able to agree. Still, considering that for the last fifty years it has not been respectable among physicists even to mention the word ether, this may perhaps indicate the beginning of a turn in the tide. There are other points also which may be more conveniently discussed in the section, now near at hand, which is to be devoted to the position of mathematics in the service of science and philosophy. But in addition to these you revert to your initial disagreement with my plea for a differentiation between the study of the universe as it is in itself and of the man-centred universe as it is for us. You insist that the second is the only sphere to which we can possibly have access. You say 'to me . . . there is no reality but experience' (p. 114). This challenge I should have liked to take up at once, for the dictum is one to which I find myself in uncompromising, and indeed vehement, opposition. But this is really the crux of our whole discussion, and it cannot be dealt with merely as an incidental part of a section in our dialogue which already touches upon a considerable variety of topics. I would prefer therefore to reserve my answer till we review in the concluding section our respective positions as a whole.

5

There are still however three specific matters which may properly come in here, since they attach to what you say about the ether question, when you ask what value would an ether hypothesis be after all, even if it were to be accepted; would it contribute enough to the advancement of knowledge to compensate for the revival of a controversy that is now dead and might well remain so, with no one any the worse. My substantial answer will be that Truth must be pursued at all costs—as you yourself are the first to emphasize when you spend yourself in an arduous controversy with the proponents of the juvenile space-travellers myth. But there are three particular matters that have long been in debate on which an ether hypothesis might present a new aspect, and I would ask leave to mention these before we go further.

The first relates to the important new conception of continual creation. It is being suggested by some physicists that the generally accepted idea that the universe is 'running down like a clock', through the continuous diffusion by the incandescent stars of an original, once for all, cosmic stock of energy, may be mistaken. It is now considered probable that this dissipation is being made good, perhaps as fast or even faster, by the continuous creation of new matter in the form of particles, or of atoms of hydrogen. Some of these, here and there, may congregate together by mutual gravitational attraction to form the various kinds of atoms and molecules. So the stars may have been evolved, and may still be evolving. This theory is of great interest, not only to science but also to the philosopher, with repercussions that will closely concern religion also.

When it is asked, however, where these pristine particles and atoms come from the answer given by the proponents of the new theory are unconvincing. Professor Hoyle for instance writes, 'From time to time people ask where the created material comes from. Well, it does not come from anywhere. Material simply appears—it is created. At one time the various items composing the material do not exist and at a later time they do. This may seem a very strange idea and I agree it is, but in science it does not matter how strange an idea may seem so long as it works—that is to say, so long as the idea can be expressed in a precise form and so long as its consequences are found to be in agreement with observation. In any case, the whole idea of creation is queer.' Professor Bondi says the same: 'It should be clearly understood that the creation here discussed is not out of radiation but out of nothing'. And the former Astronomer Royal, Sir Harold Spencer Jones, writes, 'Questions that are often asked are: What is the matter created out of, and what form does it take? It is created out of nothing: it must be supposed that there is literally a true creation going on as a continuous process.' While Professor J. R. Oppenheimer, on the other hand, says, in his Reith Lectures, that what 'physicists call vaguely, and rather helplessly, "the new particles" . . . are the greatest puzzle in today's physics'. But this answer, or absence of answer, cannot possibly be allowed to stand as part of any picture of the universe established by physics. For the idea of 'creation out of nothing' is not science at all: it is mere magic. An ether hypothesis, however, if it were accepted, would regard the production of new particles—as also of new radiations or new lines of force—as part of the normal functioning of the cosmos. The answer to the question 'where do they come from?' would be—out of the universal primordial matrix of quiescent energy, activated in those patterns.

6

My second point relates to the theory of an expanding universe, which has been generally accepted by present-day physicists, but on which an ether hypothesis may have something different to say.

First, let us recall what exactly is the astronomical discovery out of which the theory has originated. It is termed 'the red-shift in the spectra of distant nebulae'; but this term may give the mistaken impression that there is a shifting of the place of the red lines in the spectrum relatively to the other lines. That is not so: it is a question of the shifting of the spectrum as a whole relatively to the normal position of the lines as observed in the nearer stars and galaxies. This re-positioning is a movement which carries the lines of the visual part of the spectrum towards the red end. But it begins in the ultra-violet, which becomes narrower, and infringes upon what had been the violet; all the other colour lines are moved correspondingly, so that where there had been blue there is now green, and so on, until the red infringes upon the infra-red. The final result is that there is less of the ultra-violet and more of the infra-red. Our senses, however, cannot be aware of this, because neither of those bands is visible to us; but it is revealed very clearly by the spectroscope.

The possibility of such a red-shift was first predicted by the Austrian physicist Doppler rather more than a hundred years ago. Its existence was soon verified by astronomers, and it was named the Doppler effect. It was attributed correctly to the movement of a star away from the observer directly along his line of sight. It is usually compared to the change of pitch of the whistle of a railway engine when the train is approaching and then receding; and this is due to the sound-waves in the air being compressed when the train is approaching, or elongated when it is receding: the change of pitch varies in proportion to the speed of the train.

When, in the present century, the astronomers at Mt. Wilson and Mt. Palomar discovered a similar shift in the spectra of the distant nebulae, it was naturally assumed that this might also be a Doppler effect; if so, it might be a consequence of the same cause. From this it was deduced that the nebulae, or clusters of nebulae, were all rushing away from one another, and therefore from our own galaxy, the Milky Way. In the absence of any alternative explanation, this has become the standard view of present-day physics, and is taught to students in the textbooks.

But some of those who are best qualified to speak have always felt doubts about that assumption and the deduction drawn from it—Dr Edwin Hubble in particular. He was one of the workers in those great American observatories and one of the principal pioneers in this field. If we read the papers and lectures that he published we cannot fail to note how he expresses those doubts again and again. 'If red-shifts are interpreted as Doppler shifts . . . ', he says in one of them; or again, 'if red-shifts do measure the expansion of the universe . . . ' He described the theory itself as leading to conclusions that are 'rather startling', and 'strange and dubious'. And not long before his death in 1953 he wrote that the results of the exploring work that had been done were 'a definite step in the observational approach to cosmology': but—he went on to say—'the essential clue, the interpretation of red-shifts, must still be unravelled. The former sense of certainty has faded and the clue stands forth as a problem for investigation.'

Dr G. J. Whitrow, in his book *The Structure of the Universe* (1949), says the same: 'The whole problem bristles with complications . . . the ultimate verdict has not been given'. M. A. Ellison, in an article in the authoritative scientific quarterly *Endeavour* (July 1953), writes, basing himself upon Hubble, 'The red-shifts may be interpreted as indicating velocities of recession . . . Equally well, they may arise from the action of some unrecognized principle leading to the loss of photon energy in intergalactic space. Our ignorance in this field is so complete that it is wise to suspend judgment until further observations can decide the issue.' Hubble himself summed up the situation in his Oxford lectures: he speaks of 'the phenomena of red-shifts whose significance is still uncertain'. He says that 'alternative interpretations are possible . . . Red-shifts are produced either in the nebulae, where the light originates, or in the intervening space through which the light travels. If the source is in the nebulae, then red-shifts are probably velocity-shifts and the nebulae are receding. If the source lies in the intervening space, the explanation of red-shifts is unknown but the nebulae are sensibly stationary.' In that case, he says, 'they represent some unknown reaction between the light and the medium through which it travels'.

Backed by this weight of authority, we may refuse to regard this question as already settled, and in favour of the expansion theory, as most physicists seem ready to do; and we may feel free to explore the other possible alternative — that some cause, hitherto not recognized, is operating upon the light all the time and all the way, from its first emission in some distant galaxy to its reception in our terrestrial observatories. An ether hypothesis may suggest the nature and mechanism of such an alternative.

In the first place let us recognize that the phenomenon, which appears in the spectroscope as a shifting towards the red end of the spectrum, is really a general displacement of all the wave-lengths of the whole spectral band. The term red-shift, therefore, is a misnomer. We may rather speak of it as an elongation of radiation wave-lengths. This involves an equivalent lessening of their frequencies. That is to say that, during a given number of transitions from one wave to the next in a given time, there are more in the region near the emission than in the region of its end in the astronomer's observatory. And this can be explained as caused by an infinitesimally small slowing down during the transition from any one wave to the next. Those who do not accept (may I venture to say do not *yet* accept) an ether hypothesis of some sort would describe this as 'a loss of energy'. Ether partisans would say that one wave of activation is not an exact repetition of the one before; but that each in turn has suffered a very slight retardation, with an equivalent elongation. This is far too small to be observable in the early stages of the process; but it is cumulative—and in the case of the distant galaxies it may amount to as much as 20 per cent. Since it is cumulative it must vary in accordance with the distance covered—as present observation shows to be the case.

Here, once more, we find nature using the same model as the mechanism in different circumstances. An actual elongation of radiation wave-lengths has been recently observed in the research made possible by the new explorations of the Earth's upper atmosphere. It has been established that at high altitudes there is a transformation of the Sun's ultra-violet short-wave radiation into longer-wave heat radiation. This remarkable process of accumulation through repetition is also found in other quite different departments of science, and is very familiar to us in our ordinary activities. It is seen in muscular fatigue: everyone knows the difference of effort at the end of a long walk from its beginning. It is seen also in metallurgy, which has borrowed from biology the word 'fatigue' to describe the deterioration in metals through long continued use — with disastrous results sometimes in crashes of airplanes that are otherwise inexplicable. And it is most familiar of all in engineering, where it is termed friction. Every motor car owner is fully aware of the effects of friction, if long continued and if not provided against by sufficient lubrication of the moving parts.

So we may conceive the successive pulses of a train of expanding spherical waves of ether activation, each one suffering an infinitesimal retardation and elongation, as a feature of the effort of transmission to the wave that will be next in the layer of quiescent energy ahead of it. The effect is similar to muscular or metallic fatigue, or friction in machines. Cumulative in all cases, it is very slight at the beginning but increases with repetition; and it is measurable in exact proportion to the number of repetitions, which is itself proportionate, in cases of movement, to the distance travelled.

Here, then, may be the 'hitherto unrecognized principle' which affects the light radiation during its journey, and which might be admitted as a possible alternative to the notion of a universal recession of galaxies.

Two minor points remain to be mentioned. We are usually told that what we are seeing in the spectroscope is the star or the galaxy as it was at the time when the light radiation started on its journey—perhaps millions of years ago; possibly meanwhile the star or galaxy may even have disappeared in some cataclysm. That is no doubt true of certain particulars—chemical composition, atmosphere and others. But it is also true that what we are actually looking at is the spectrum arranged as it is now at the time of reception, and not as it was at the distant time of its origin.

The other point is the effect of all this upon Doppler's original discovery. It seems clear that it has no effect at all. Wherever a star or a galaxy is in fact in motion away from us in our direct line of vision, the so-called red-shift will appear. Whenever this happens it will still continue to happen, unaffected by the acceptance, or the possible supersession, of the speculation that the red-shifts of the distant galaxies are of the same kind, but much smaller in magnitude for a given distance.

Our alternative, then, may claim to conform to all the conditions that are imposed by the results of observation. First, no doubt is thrown upon the reality of the Doppler effect in the cases where it has been directly discovered and verified. Second, the ether process applies to all stars and galaxies, whether relatively near to us or distant; but it is on so infinitesimally small a scale that it is far beyond the limits of perception in our telescopes and spectroscopes, and it has only been discovered at all because—as in other cases—it is cumulative with repetition. The measurement of the effect must therefore conform with the measurement of distance—the greater the distance the greater is the effect. But this is not because the galaxy that is being observed is moving in recession from us at a greater velocity than the nearer stars or galaxies, but because the wave-lengths that are recorded in the terrestrial observatory have all been elongated as part of the process of transmission; and the longer has been the journey the greater therefore is the lengthening.

7

My third point is connected with the long-sustained and still unsettled controversy on the wave-particle relation. In certain experiments in the laboratory, effects are produced which, the physicists hold, have clearly been caused by electrons or other particles, and not by waves, yet, at a later stage of the same experiment there are phenomena which are typical wave effects. This has given rise to much perplexity, and to some strange speculations. One of these is that every wave possesses some particle characteristics. Another is that they are all some kind of hybrid, which has been nicknamed 'a wavicle'. Or again it is suggested that it may be a question of 'aspects'—we see what is happening from one line of approach, and then we see something different from another line of approach: the difference is not objective, in the phenomena, but subjective, in the observer. At the end many—and apparently the great majority—of physicists seem to have come to the conclusion that the puzzle is insoluble, and prefer not to trouble themselves about it any more: they prefer to translate the situation into mathematical symbols, and to cope with it by means of differential equations, that have become ever more complicated and artificial.

Here again an ether hypothesis could offer a possible explanation along other lines. Regarding both waves and particles as patterns of activated energy, it may conceive that in certain forms they are easily transmuted from one into the other. Sometimes indeed the particles may be enduring for long periods, or indefinitely—either as free units, electrons for example, or positrons, or neutrons, or when combined into atoms. We know what tremendous forces have to be employed to 'split the atom'. But other particles or atoms are quite unstable—like the newly discovered transuranium elements, which may have a life limited to a few hours, or even seconds, or fractions of a second. We may conceive it possible that there are conditions in which certain kinds of particles may transform themselves into other kinds, or into waves; and certain kinds of waves into waves of other kinds, or into particles.

Such mutations are in fact frequently being brought about in the laboratory or in nature. An experimenter may pass an electric current—which is held to be a stream of particles—electrons—into a cathode tube, and get out at the other end a sequence of waves—the electromagnetic Röntgen waves that we call X-rays.

Again, it is our familiar experience that heat is produced in two different forms—either radiant heat, which is transmitted by waves (having their own place, beyond the infra-red, in the Clerk Maxwell wave-band); or molecular heat, transmitted by some material substance—solid, liquid or gaseous—which is in a state of thermal agitation. We can see the first in the heat from an electric bulb or electric heater, or a coal fire: we can see the other in the radiators of a domestic hot water system; or in a heated laundry iron, for example; or in the atoms or molecules of the atmosphere on a hot day. May it not be the same with the transmission of light?

Many scientists would say, however, that we have to choose between the two—it is either a wave process or a photon process. But why should we consent to be so restricted? Why should we not conceive that nature uses both processes, with light, as it is recognized she does with heat? And, further, that either can be transmuted into the other? The molecular heat produced by the combustion of a mixture of coal and coke in the small furnace in the basement of my house, becomes molecular heat in the radiator by my side, and then the radiant heat, which is at this moment warming the air of my room, and myself with it. Why should there not be similar transformations with light?

Need the astronomer be compelled to believe either that the light rays which enter his telescope and spectroscope have travelled across the universe in every direction and perhaps for millions of years, as streams of particles—a very improbable hypothesis: or else that electromagnetic waves are able to affect the grains of chloride of silver in a photographic film so as to give a picture of a particular section of the heavens — also improbable?

Nature evolved the lens millions of years before man invented it. Originally a localized thickening of the skin of primeval fishes, it evolved into the highly complicated visual organs of animals of all kinds, man included. It would appear that this light, combining a number of waves of differing wave-lengths to produce what we call a white light, with separate effects that we call colours, strikes the eye as waves: but is converted by the lens into particles, which are capable of stimulating the nerve-endings at the back of the retina and so of bringing the relevant portion of the brain into action.

Similarly with the experience of the astronomer, the light from the stars and the galaxies has travelled across the universe as a pattern of expanding spherical waves—the 'moving configuration'—until a segment of it is caught by the lens of the telescope. There it is subjected to the process of transmutation, the mechanism of which is not yet known. It emerges from the lens as particles, as photons. So that when we are asked, which are right, the advocates of the wave theory or the advocates of the photon theory?—the answer may well be (as so often with scientific or philosophic dilemmas): both are right, but they are to be viewed as a succession of two phenomena, and not as a single phenomenon.

In the book already cited I have dealt more fully with this speculation and it is not necessary that I should repeat it here. I would merely submit that perhaps the solution of a puzzle, which is being regarded as insoluble, may be found in the principle of transmutation. If experiment and observation have not confirmed that speculation, that may be, not necessarily because it is wrong, but because science has not yet produced another Hertz or Röntgen to discover the missing link.

The genius of Newton seems to have foreseen all this; for in his *Opticks* he discusses 'the changing of Bodies into Light, and Light into Bodies', which, he says 'is very conformable to the Course of Nature, which seems delighted with Transmutations'.

PROFESSOR DINGLE:

The three points you raise are of great interest, and serve excellently to illustrate the divergence of your approach to physical problems from that of many mathematical physicists today. I should like to comment briefly on each of them before turning to the wider question.

1

There are two elements in the hypothesis of continual creation which I think it is important to distinguish, but which, in fact, are not distinguished at all by the proponents of the idea. There is first of all the possibility that in fact particles suddenly appear at places where there were none before, without having come from other places. The new particles can then be said to have been 'created' at the moments of their appearance, provided that by that word we do not indicate anything more than the fact of their so appearing. The second element of the hypothesis is the positive assertion that they have come to exist 'out of nothing'.

The first possibility is perfectly legitimate as a scientific hypothesis if there is evidence for it or if, by assuming it, we can draw deductions which can be compared with observation. In the present instance there is no evidence at all, but by combining the hypothesis with certain general ideas which it is customary to employ in cosmology, with certain facts of observation, and, I fear, with certain arbitrary 'principles' of the same character as the Aristotelian dogma that all celestial motions must be in perfect circles, it is possible to draw conclusions as to what should be observed in certain prescribed circumstances. This is being done, but whether the results are sufficiently in agreement with fact to allow the hypothesis to survive is at present an open question.[1]

It should not be forgotten that in cosmology the proportion of observable to relevant facts is extremely small, and the interpretation even of the observable facts is very hazardous. No one who knows how, in laboratory work, where one can experiment freely and measure with almost incredible accuracy, quite false conclusions can still be drawn (witness the false results expected of the Michelson-Morley experiment for instance), can regard cosmological speculations, whatever they may be, as having much probability of approximating to the truth. We must, of course, make them because it is the best we can do, but there is a tendency among theoretical scientists to assume that if you can construct a theory that accounts for all the facts available, its chance of being correct is independent of the number of necessary facts that you do not know. This tendency is particularly evident among cosmologists, and it should always be borne in mind when reading their confident pronouncements.

As I say, this is no reason for suppressing any speculation that may point the way to the acquisition of new knowledge; it is only a warning against supposing that the speculation is true. After all, there seem to be only three possibilities concerning the origin of the material universe: it may have existed from eternity; it may have been created all at once a finite time ago; and it may have been created piecemeal at different times. In the

---

[1] Since this was written observations by M. Ryle and his collaborators have been announced which show that the predictions of the hypothesis are not in agreement with fact.

absence of knowledge the hypothesis of continual creation has thus a chance of 1 in 3 of being correct. It is not unreasonable, then, to attempt to give it a precise form and explore its possibilities as well as we can. But, by the same token, there are only two possibilities for any hypothesis: it may be right and it may be wrong. I do not think we should be too much influenced in our scientific work by the deduction that every hypothesis has equal chances of being right and being wrong.

But it is the second aspect of the hypothesis to which you have drawn attention, namely, that the particles are said to be created 'out of nothing'. That is, as you rightly imply, the abnegation of the scientific adventure altogether. What, after all, is science? It is the attempt to link all our experience together into a rational system, to see each phenomenon not as an isolated happening but as related in some way to other phenomena. When we say a particle is created 'out of nothing', then, we are merely saying, 'Here is a phenomenon which I will not attempt to relate to other phenomena'. If that is to be permitted we might as well (apart from utilitarian motives) have done it at the beginning, and said of thunderstorms, for instance, that the sound came 'out of nothing'. Why go so far as we have done, only to give up when we meet with a new obstacle?

The reason is, as so often, that the authors of this hypothesis are symbol-manipulators, who have forgotten that their work does not become scientific until the symbols are validly related to experience. Their equations can be satisfied if symbols which have a non-zero value for a certain position of the time-co-ordinate have a zero value for a position to the left of that, and, materializing the symbols—the only reality they acknowledge—they say, 'something has come out of nothing'.

I therefore regard your idea of active and quiescent energy as infinitely preferable to the abandonment of science, but it seems to me that the possibility of this spontaneous appearance of particles is so speculative at present that it is hardly worth while asking whether it can be interpreted in detail as a transition from quiescent to active energy. As I have indicated before, I think this latter idea must be given a more precise and detailed form before physicists can make use of it, and if that is done it should be done by consideration of well-established facts, of which there is no shortage, rather than such speculations as this. But if, in the future, we should find reason to suppose that continual creation actually occurs, then I think some such conception as yours, in one form or another, will be inevitable. Certainly an existing particle is a quantity of energy according to present ideas, and if it is to be related to the circumstances in which it appeared, the particular feature of those circumstances that is linked to it must be identified. In the light of present knowledge the probability that it will be expressible as a form of energy seems very high.

2

Your claim that there is no conclusive reason for supposing that the nebular red-shifts are Doppler effects is perfectly justified. I think that most, if not all, astronomers would agree with this, though probably the majority would express a preference for the view that they are. I think this is bound up with the fact that the general theory of relativity lends itself naturally (though not inevitably) to an interpretation of the universe in which continual expansion is an essential feature, and the temptation to choose an interpretation that admits of mathematical development is very great. There is also the fact that we have observational evidence that expansion could cause a red-shift, whereas if the universe is not expanding, we have no definite idea at present of what the cause might be.

But the Doppler effect itself is by no means so simple a matter as is usually supposed. It was introduced into physics, as you say, by Doppler as a necessary consequence of the nature of wave motion, and regarded as applying in exactly the same manner to sound and to light waves. But, so regarded, the effect is different when the source moves to or from the observer from what it is when the observer moves to or from the source. In the former case the waves themselves are changed—compressed or drawn out, as the case may be—but in the latter case nothing happens to the waves; the 'frequency' that is changed is that of the observer's encounter with them. As a result, the amount of the effect differs (though very slightly unless the velocity is extremely great) in the two cases.

Since Doppler's time we have obtained direct evidence for both a moving observer (in the orbital motion of the Earth) and a moving source (a component of a double star) that the effect exists for light, but we cannot measure accurately enough to determine whether both effects are the same or not. According to the postulate of relativity there should be no difference, since it is impossible to ascribe any such motion to one of the bodies concerned rather than the other, and the formula giving the amount of the displacement which is now generally accepted accords with this. But this raises problems that have not been sufficiently considered.

Suppose we have three observers, A, B and C, in line, with B between the others. Let them all emit light of the same frequency. When they are all relatively at rest, each will observe the spectrum of the light from the others to be identical with the spectrum of his own light. Now let B move rapidly towards A. Then we believe that he will see A's light displaced towards the violet and C's towards the red, as compared with his own. Now this cannot be due to a change in his own light, for that would make both of the others appear to be displaced in the same direction. Nor can we suppose that a moving force applied to B can make two undisturbed sources change their light in opposite senses. So we have changes in the relative frequencies of three sources of light when none of them has itself changed.

The only possible way of explaining this is by recognizing that the 'frequency' (or 'wave-length' or whatever it is that the spectroscope measures) of light is not a property of the light but of a relation between light and an observer. This, be it noted, does not depend on any theory of light at all; whether we are dealing with waves or particles or anything else at present inconceivable, this conclusion holds good. It is therefore useless to try to explain the Doppler effect in terms of a particular theory of light. Rather our theories of light must themselves conform to this condition, that what the spectroscope reveals is not an intrinsic property of light but a property of its relation to the spectroscope; it changes when that relation changes. Mathematicians usually write as though the Doppler effect could be deduced from the special theory of relativity. Actually we must suppose it to exist before that theory can be applied to it. The special theory of relativity is concerned only with the *velocity* of light; the frequency (whatever that may be) has to be discovered by experiment, and the relativity theory would be exactly the same if it did not exist.

The relevance of this to the problem of the nebular red-shift is this, that if that is truly a Doppler effect—an effect of relative motion — then no change in the light itself is valid as an explanation, and no means that our ingenuity might devise of examining the light independently of its relation to ourselves and the nebulae, would be expected to reveal any difference in the light from one nebula and that from another more distant and more rapidly moving one. On the other hand, if your theory of fatigue is correct, such an examination would show that, in properly comparable samples, the latter would carry less energy than the former. I confess that I can see no way in which such a test could be made. If it were established in the laboratory that Einstein's second postulate is invalid and that light from relatively moving sources travels through space with different velocities (see p. 66), then a test would be possible. What is called 'the aberrational constant' (an apparent displacement of a distant source of light associated with the Earth's orbital motion) would vary with the velocity of the distant source, and a determination of the 'constant' for different nebulae would therefore reveal whether they were moving or not. But this laboratory measurement has not been made, and if Einstein's theory is correct the aberrational constant should show no such variation. At present, therefore, we can only say that the red-shift may be due to expansion or to some kind of fatigue of light, and await further knowledge.

There is, however, this to be added, that if travelling light loses energy by fatigue, one who regarded energy as something that is conserved would expect the lost energy to appear in some other form. Friction in ordinary materials produces heat, which is measurable and is proportionate to the kinetic energy lost. Your lost energy would presumably be 'quiescent', but in the absence of any means of verifying this it would be difficult to distinguish such a hypothesis from one in which the energy was destroyed altogether. That would seem to be no better than creation out of nothing—continual annihilation of something. Those who profess to believe in creation out of nothing could not, of course, use this argument against you, but genuine scientists would be free to do so. You might deny annihilation and predict that one day a means will be found of detecting the energy apparently lost, or it would be open to you to postulate that this energy reappears in spontaneously created particles. This is, in fact, what you have suggested in the section already discussed (p. 120).

Let us suppose that you make that hypothesis. Then you would differ from the advocates of the 'steady state' theory, as the existing theory of 'creation out of nothing' is called, in that you would assume that particles appear spontaneously in a static universe, whereas they assume that an amount of existing matter equivalent to that newly created is carried beyond possibility of observation by 'the expansion of the universe'. The one-way process in the universe, according to your theory, would be the conversion of radiation into matter. As time goes on, more and more atoms appear and radiation becomes less and less energetic. This, of course, would have to face the problem attending all one-way theories, namely, that of extrapolation backwards and forwards in time to a beginning and end of the universe. The 'steady state' theory claims to escape this problem since its universe as a whole remains eternally the same, but this is only an apparent escape. Each galaxy had to have a beginning, it is known to exhibit one-way processes, and since it will not pass out of the observation of its own inhabitants the problem of its future must be faced.

If I had to choose between the two I would choose your theory, but I retain the view that the whole subject is at present so highly speculative that no opinion on the matter is worth holding. We want more knowledge, and although it will be a long time before we shall get all we need for a really well-founded theory, there is much that may soon be available. A decision concerning the proper interpretation of the red-shift is what we most require, and this may well be not long delayed. In the meantime I think it is best to turn to problems offering better prospects of progress.

3

The wave-particle problem need not detain us long. It is indeed a real problem, not to be solved by merely showing that two sets of mathematical processes stand in a one-to-one correspondence with each other, but it is not a simple matter of light (or electrons too for that matter) appearing sometimes as waves and sometimes as particles; it presents both aspects *at the same time*. Take, for instance, the spectrum of white light formed by a diffraction grating. We can explain the formation of the spectrum only by supposing the light to consist of waves, but we can explain the distribution of energy among the different colours only by supposing the light to consist of particles. This difficulty cannot be overcome by any transformation hypothesis.

LORD SAMUEL:

I have a few things to say on your comments on my three points, and am glad to note that there is a measure of agreement on each of them.

1

Continual creation. You are definitely of the opinion that the idea of 'creation out of nothing' is unscientific and inadmissible. As Oppenheimer says, the present prevailing opinion among physicists regards the problem 'helplessly', and leaves it as 'the greatest puzzle in today's physics'. But our opponents may legitimately ask what alternative we have to offer. To this, for my own part, I would answer that, if an ether hypothesis is regarded as worthy of consideration on its own merits, it might provide, among other things, a clue to this puzzle of continual creation. It would then, on the one hand, bring about a simplification of that issue, and, on the other hand, might make the ether hypothesis more worth while. But whether or not it is to be accepted as sound is a wider matter for physicists themselves to decide, and not for philosophers or other laymen.

2

Theory of an Expanding Universe. Here you agree that my contention that 'there is no conclusive reason for supposing that the nebular red-shifts are Doppler effects' is perfectly justified; and you add that you 'think that most, if not all, astronomers would agree with this, though probably the majority would express a preference for the view that they are'. At the present time I cannot expect more than such a suspensory judgment. This matter also is one for the specialists. The important point now is to secure, first a general recognition that an expanding universe is not already a settled issue. Afterwards the results of additional research may tend to show that an alternative to nebular recession may be a hypothesis of cumulative elongation of the spectra of electromagnetic wave-lengths throughout the whole length of their transmission. Indeed the most recent observations yielded by the *sputnik* satellites record that such elongation is a process now happening in the universe in the regions beyond the limits of the Earth's atmosphere and range of gravity, in the conversion of the Sun's ultra-violet rays into heat rays.

3

Wave-particles. This, you agree, is a real problem, but you do not agree with my speculation that it is an instance of a normal ether process, the transmutation of one pattern of energy into another. You cite (p. 133) a particular experiment in which this cannot hold. But it does not appear that this is universally regarded as conclusive. Professor Louis de Broglie has long been regarded as one of the most eminent physicists of France. In a recent book *Physics and Metaphysics*[1] he considers this question at some length. He writes:

'In principle there is nothing against the view that energy, while always conserving itself, can pass from the material to the luminous form and vice versa. We know today that it is actually so . . . This final union of the conceptions of light and of matter in the unity of this protoform entity which is energy, has been completely proved by the progress of contemporary physics on the day it discovered that material particles are capable of disappearing while giving rise to radiation, whilst radiation is capable of condensing into matter and of creating new particles . . . All these facts clearly prove that light and matter are only different aspects of energy which can take in succession one or the other of these two appearances . . . Finally, light has just revealed itself to us as capable of condensing into matter, whilst matter is capable of dissipating into light.'

He says again, 'The wave aspect asserts itself only when the corpuscular aspect vanishes and vice-versa . . . As though by a curious precaution of nature, the two aspects—the corpuscular and the wave—play a sort of hide-and-seek game, so that they never come to oppose one another.'

One may ask why de Broglie does not draw the conclusion that the case is one of succession and not of simultaneity; again one of transmutation, of disappearance and substitution and not one of strained metaphysical paradoxes—a single entity being at one and the same time a spherical expanding electromagnetic-wave and also a discrete localized particle.

4

So that, when you ask me what advantage an ether hypothesis

---

[1] Published by Hutchinson, 1955, see pp. 67-8, 138-9, 240.

would be, even if it were to be generally accepted by physicists, I would still submit those three examples.

1. The appearance of new particles, and perhaps even of new atoms, would be seen as normal activations of quiescent ether.

2. The so-called red-shift of nebular spectra would be discovered to be nothing else than a general elongation of electromagnetic wave-lengths as an equally normal feature of wave transmission: far below the level of detection over relatively short distances, but cumulative, and easily observable over greater distances; the degree of elongation being therefore proportionate to the distance travelled.

3. The puzzles that arise in wave-particle relations may, after all, perhaps be revealed to be one more example of nature's frequent model of the transmutation of entities. If at one stage in an experiment the researcher finds himself dealing with wave-effects, and at another with particle-effects, a conceivable explanation might be that he is in fact dealing at one time with waves and at another time with particles.

III

SCIENCE AND MATHEMATICS

PROFESSOR DINGLE:

1

Implicit in much of our discussion so far is the question of the place which mathematics occupies, or should occupy, in science. I know that to you, as to me, this is a matter of great importance. It is obvious, of course, that modern physical science is mathematical through and through, and whether or not Galileo was justified in saying that the book of nature is written in the mathematical language, there is no question that that is the language of the modern description of nature, at least so far as the subject-matter of physics is concerned. Nor is it possible at this time of day to see this as anything but inevitable. It would be as idle to challenge the mathematicizing of physics as to rail at poets and orators for speaking the language of imagery. Yet there is in this a danger. A famous statesman was once charged with being 'inebriated with the exuberance of his own verbosity'. Whether or not this was a slander you will know far better than I, but it does prompt the question whether mathematical physicists might not sometimes be inebriated with the exuberance of their own symbolosity, if such a word may be coined. My answer would be decidedly in the affirmative, and this is all the more serious because of the absence of effective criticism.

I expect you know 'the famous story of Euler [the great eighteenth-century mathematician] and the atheistic (or perhaps only pantheistic) French philosopher Denis Diderot (1713-84)'. I tell it in the words of E. T. Bell.[1] 'Invited by Catherine the Great [of Russia] to visit her Court, Diderot earned his keep by trying to convert the courtiers to atheism. Fed up, Catherine commissioned Euler to muzzle the windy philosopher. This was easy because all mathematics was Chinese to Diderot. De Morgan

[1] E. T. Bell, *Men of Mathematics*, Chap. 9 (Pelican Books).

tells what happened (in his classic *Budget of Paradoxes*, 1872):
"Diderot was informed that a learned mathematician was in possession of an algebraic demonstration of the existence of God, and would give it before all the Court, if he desired to hear it. Diderot gladly consented . . . Euler advanced toward Diderot, and said gravely, and in a tone of perfect conviction:

'Sir, $\dfrac{a + b^n}{n} = x$, *hence God exists*; reply!'" It sounded like sense to Diderot. Humiliated by the unrestrained laughter which greeted his embarrassed silence, the poor man asked Catherine's permission to return at once to France. She graciously gave it.'

Now I do not suggest that modern mathematicians silence criticism with the same intentions as Euler: indeed, I think—without denying that there are exceptions—that, in a sense, they combine Euler and Diderot in their own persons, and honestly believe what, if they possessed normal reasoning powers or normal persons understood the mathematical language, would at once be seen by either to be absurd. But the potential consequences are indeed very serious, and I think it is necessary to try to understand how mathematics entered science, in order to understand as well as we can what its proper function there—quite apart, of course, from its status as a body of abstract reasoning independent of experience—may be.

2

The fundamental problem of all philosophy, including science, is, as I see it, to understand our experience, to see it as a rationally connected system in which each element is related to every other, instead of as a succession of apparently independent and unrelated happenings as it first presents itself. No matter whether we express our problem as that of understanding the universe, or discovering why things are as they are, or in any other way, it all comes down to the description I have given because all our knowledge of anything at all comes to us through experience—in the most general sense, of course, not sense experience alone.

Men naturally began with the most obvious relations—night succeeds day regularly, hunger may be removed by eating, and so on—and, taking this as a basis, tried to extend the system of relations over all experience. Insurmountable obstacles appeared, however, and at last, in the seventeenth century, a fresh start was made *ab initio*. Natural experience was ignored, and the new start was made with *artificial* experiences—balls made to roll down grooved slopes and such things, that would never have come to pass had they not been specially contrived. Not only so, but the aspects of these artificial experiences that were considered were only those that could be *measured*—the distance covered, the time of rolling. Thus was set the pattern that physics has followed ever since.

The advantage of this procedure was twofold. First, these artificial experiences were more controllable and precisely definable than the things that occurred naturally, and they could be made to recur with differences that, if they existed at all, were inappreciable. Consequently, any relations found between them had a general and permanent significance. Secondly, by measuring—i.e. by obtaining numbers to represent the artificial experiences—it was possible to use the most powerful tool of reasoning available, namely, mathematics. Thus the general problem of giving a rational account of all experience was reduced to the particular one of giving a mathematical account of artificially produced experience.

The attack on this particular problem is what we now call physical science. It is not the whole of science, still less the whole of philosophy. There are branches of science — biology, for example—in which the manufactured experiences are not susceptible of measurement but are still amenable to other kinds of reasoning, and others—e.g. some departments of psychology—in which the interference with nature is very slight indeed. Naturally, the more we limit our endeavour the more progress we make, so that physics is a more advanced science than biology and biology than psychology. But they are all parts of the one general problem—that of uniting all experience into a single rational system.

Let us, however, continue with physics. The next step, after the substitution of measurement for direct response to nature, was the substitution of precisely defined concepts for the measurable experiences. I will give two examples to show what I mean. Take first any ordinary object, say a stone. For our direct experience of the stone physics substituted its weight—the number we get when we perform a particular operation with it. Its colour, shape, hardness, etc., were all put aside, and only the result of this particular operation was taken to represent it in the sub-science of mechanics. But very soon this artificial experience gave place to its *mass*, which, as I shall show directly, is not an experience at all, even an artificial one, but a mental creation which indeed is not unrelated to weight, but enables simpler mathematical expressions to be constructed than would be possible if we had to limit ourselves to the unadulterated measurements themselves. In order to maintain contact with experience, mass had, of course, to be defined in terms of the artificial experiences, and this was done, but the mathematical equations, which express the relations found in experience, did so only indirectly.

As a second example, take temperature. As a natural experience this relates to the fluctuation of hotness and coldness that we know as summer and winter succeed one another, for example. Physics substituted for this the artificial experience of observing the height reached by a liquid (usually mercury) in a tube. This artificial experience was in turn displaced by a concept too complex to be described here in detail, which may be expressed as the mean kinetic energy per molecule of the body concerned. This is a pure concept, but, like mass, it could be precisely related to our artificial experiences of measurement, and so our equations involving it do ultimately express relations between experiences, though in a highly disguised form.

These examples are typical of the whole of physics. Take any physical equation you like, and every symbol occurring in it is far removed from experience, yet, if it is a valid equation, ultimately expressible in terms of experiences, though only artificial ones. It is finally reducible to the relations between things that occur in laboratories but would never happen of their own accord.

It may well be asked: if this is so, how is it that the achievements of physics can be turned to such direct account in our lives? The answer is: because there is a purely empirical correspondence between the symbols and our *natural* experiences. I emphasize that this correspondence is empirical: we do not in the least understand it and we could not have predicted it, but it happens to exist. When the symbol for mass turns out to represent a large number we happen to see some big piece of matter. When the symbol for temperature represents a big number we happen to feel hot. Why this is so we have not the slightest idea:

for all our sophisticated knowledge of physics we depend in the last resort on what we happen to find.

This can best be seen by considering cases in which the correspondence breaks down. The law of gravitation tells us how a mass-point moves in a gravitational field. The ordinary person thinks it tells us how a planet moves round the Sun, and so it usually does, but only because of the empirical relation I have spoken of. In the case of the Earth-Moon system it happens that the mass-point in question is situated inside the Earth, though far from its centre, but if the Moon had been slightly farther away the mass-point would have been in empty space, the point which we call the *centre of gravity* of the system. The mass at that point would be represented by a large number, but we should see nothing there. If the whole system were smashed to smithereens by internal forces, the planet would no longer exist as such, but the mass-point would go on its orbit round the Sun exactly as though nothing had happened. The law of gravitation tells us absolutely nothing about the internal characteristics of the system: the fact that from it we can predict where *we* shall be in the orbit this time next year depends on the purely empirical expectation that things quite outside the scope of the law of gravitation will go on as before.

It is exactly the same with the other concepts. According to present views the temperature of interstellar space is far higher than that of molten lead, but a thermometer placed there, if it were capable of recording anything, would record hundreds of degrees below freezing point, and there can be little doubt that a man would very soon die of cold in such circumstances. The normal correspondence between molecular kinetic energy and feelings of hot and cold would break down, and the fact that in ordinary circumstances it *is* normal is simply something that we must accept and not pretend that we understand in the least degree.

3

We are now in a position, I think, to take a clearer view of those features of the present situation that perturb us both. A young student with a flair for mathematics is attracted to physics, where he sees scope for his talents. He is introduced to physical terms such as *mass* and *temperature*, which he has known almost from babyhood as standing for common, everyday experiences, and when he finds symbols in his equations called by the same names he identifies them and supposes that they stand directly for the experiences he knows. Nobody corrects him, for the philosophy (i.e. the understanding) of science has no place in our educational system, and so he grows up taking it for granted that nature—i.e. experience—is controlled by his equations. In the next generation he and his fellows become the leaders of physical science. They do admirable mathematical work, and their ingenuity in devising concepts susceptible of mathematical development is one of the wonders of our time, but of the relation of their work to experience — even to the artificial experiences of measurement — they have the crudest notions. Of the devious and partly uncharted road by which mathematical symbol is related to experience they know nothing. There is an occasional exception, of course, but I am describing the normal course of events and the normal result.

Errors concerning the relation of symbols to measurements, as distinct from natural experiences, are usually corrected by experiment. The mathematician predicts something, the experimenter finds that it does not happen, and the mathematician is then forced to try again. This is the traditional course of scientific progress, and it serves its purpose so long as experiment can keep in close touch with theory. In the past it has generally done so: the kinetic theory of gases, for example, which is the branch of physics in which temperature is defined in terms of molecular kinetic energy, developed as a laboratory subject, and so no one believes that the interstellar temperature of more than $1,000°$ means that a thermometer would measure that temperature: it is understood that the thermometer responds not exclusively to the mean kinetic energy of the molecules but to that combined with the number of molecules possessing that energy, and in interstellar space these are abnormally few.

But when theory has to develop without adequate experimental control—as it has in cosmology and in problems of high-speed space travel, for example—the blindness of the mathematician to the relation between his concepts and measurements is seen in its stark nakedness, if the metaphor may be allowed. His reasoning power has emptied itself into his mathematics; he has none left for the relation of the mathematics to measurements. He lives in a microcosm of abstract concepts connected by mathematical equations, like a ship that has lost its attachment to the shore of experience and has drifted off, carrying with it a crew unaware that there is anything in the universe outside their vessel. They create concepts, not because they bear a significant relation, however remote, to experience, but because they are capable of mathematical development.[1] They develop them magnificently and interpret them nonsensically. The example I have already given, of 'asymmetrical ageing', is sufficient to show this. It is not difficult to find others.

A legend has grown up that the scientist is in some sense a being superior to the rest of mankind, more dedicated, incorruptible; and the marvellous successes of science are ascribed to this loyalty which, seen fitfully in others, is in him unfailing. It is a complete mistake. The scientist is an ordinary human being, on the average no more and no less subject to the frailties of human nature than the lawyer, the politician, the economist, the historian. It is true that a superficial reading of the history of science will appear to support the illusion. There is no dearth of instances in which cherished theories have been abandoned because of some unexpected and unwelcome observation. But a deeper reading shows that this is because of the very nature of science itself. Theories are systematic descriptions of observations, and when observations are plainly seen to contradict them the scientist gives them up just as the batsman walks out when his wicket is seen lying on the ground. And until recent times, scientific theories have almost always been such that some experimental test is quickly available. When it is perforce delayed, then appeal must be made to the umpire, reason. And in those circumstances the scientist is usually less faithful to the spirit of the game than the cricketer. The batsman invariably retires when the umpire signals l.b.w., even though he may think the decision unjust; but in science, when experiment is not immediately possible an Ohm may be calumniated or a Waterston driven to suicide, not because their theories are irrational but because they are unorthodox. True, in such cases amends are made when at last experiment pronounces its verdict—an eventuality that has no parallel in cricket—but the lesson is never

---

[1] 'One of the weightiest objections brought against the hypothesis . . . is that it has not proved possible, so far, to put it into a mathematical theory'— H. Bondi, *Cosmology*, p. 32.

learnt because the history of science is little known and less esteemed among scientists.

Until recently, as I say, the verdict of experiment has usually been at hand almost immediately, but now times have changed. We have reached a point at which the necessary observations require us to have been present at the Creation or to measure something far beyond the power of our instruments or to survey the universe from a distant nebula or something of that kind. In these circumstances, what is demanded of science is, first, that its theories shall be based on the facts that we do know, and second, that they shall not require the possibility of experiences that are essentially incompatible. It is a new situation, and scientists are not prepared for it. Theories are advanced which *originate* not in facts but in assumptions (dignified as 'Principles') which are of such a nature that observation could test them if the practical means were available. On this precarious basis huge mathematical structures are built, and if the implications of the mathematics, when translated into the corresponding observations, turn out to contradict what we *do* know, that is regarded as of no importance so long as they have been correctly deduced from the arbitrary premises. Such is present-day mathematical physics.

Non-mathematical physicists, who have maintained a closer contact with experience—and they are far more numerous than the popular view of scientific activity would lead one to suppose—see this with misgiving, but they distrust their power to challenge the mathematical details, and hold their peace. Thus we have a state of affairs in which irrational mystification rides roughshod over plain reason, and the conquest is extended to the non-scientific general public who, seeing the miracles that 'science' has achieved in the practical realm, are ready to believe anything that the 'scientist' tells them, no matter how absurd it may appear. Thus is created a most dangerous state of mind, against which I protested as long ago as 1934, and again in 1937,[1] with small, if any, effect. The particular obscurantisms then rampant have now mostly gone to their own place, from which they should never have emerged, but the paralysis of the public intelligence has progressed as the hydra has grown new heads. Such is the position with which we have to deal.

[1] *Nature*, June 2nd, 1934; May 8th, 1937; June 12th, 1937.

4

The conclusion to which, as it seems to me, we are forced—that mathematical physicists, generally speaking, are deficient in reasoning power — is so paradoxical as to seem unworthy of serious consideration. But I mean it seriously. Against what may seem all the probabilities, facts compel us to acknowledge that it is true. I am no psychologist, and any attempt of mine to account for such an apparent anomaly can carry little weight, yet I cannot help thinking that the nature of mathematics, as expressed by two of the leading authorities on this subject, does enable us to understand in some measure why a training in mathematics has a tendency (not, of course, necessarily irresistible) to blunt the edge of one's sensitiveness to the realities of experience.

In an oft-quoted passage, which I am not able at the moment to quote in the original words, Bertrand Russell described mathematics as a subject in which we do not know what we are talking about or whether what we are saying is true. It is an accurate description. Mathematics consists of deductions from propositions which are formulated in the first instance quite arbitrarily—*axioms*, they are rather misleadingly called. From these, which may be few in number, together with a few rules of procedure, it is often possible to infer an enormous corpus of theorems which follow by logical necessity. The best-known example is perhaps Euclidean geometry or simple arithmetic, but these are merely two of an indefinitely large number of such systems of thought. Let us, however, take arithmetic as an example.

In themselves, the rules of arithmetic tell you about the behaviour of numbers, which are arbitrarily defined—how they are defined does not matter for our present purpose—as having certain properties. From these properties certain relations necessarily follow. For instance, the numbers '1' and '2' and the process of 'addition' having been defined, it follows inescapably that $1+1=2$. But that tells you absolutely nothing about the world of experience, because '1' and '2' are not things which we observe, but inventions of our own. We do find, in fact, that if we voluntarily identify '1' and '2' and other numbers with the results we get when we count natural objects, then in many cases the identification holds good throughout the whole of our arithmetic. This is so, for instance, with objects such as apples. If we add 1 apple to 1 apple we get 2 apples; 12 apples added to 7 apples make 19 apples, just as in our arithmetic $12+7=19$; and so on. We then consider ourselves justified in taking it for granted that, say, 10,000 apples can be arranged in a square in which each side contains 100 apples, and we should regard it as superfluous to make an experiment in order to verify this.

But this is by no means always the case. If we add 1 drop of water to 1 drop of water we get, not 2 drops of water, but 1 larger drop. If we add 1 litre of a gas to 1 litre of another gas, the volume of the resulting gas (if the result is still a gas) will depend on the gases we are dealing with. If we add 1 unit of velocity to 1 unit of velocity we get, according to the generally accepted theory, rather less than 2 units of velocity. There is no way at all of knowing to what natural phenomena our purely mental mathematical system will apply except by trial. Having found compatibility at a few points, then we can use the logical necessity inherent in the mathematical system to infer it at others, and that is why mathematics is so useful in physics and in everyday life; but that the initial correspondence will hold good is a matter of which our mathematics by itself tells us nothing. So long as we remain in the realm of pure mathematics we do not know what we are talking about when we say that $1+1=2$: only when we appeal to experience do we discover that we may be talking about apples but certainly not about ideas. And if we make an arithmetical statement about some unspecified entities, we have no idea at all whether what we say is true or not.

Now the mathematical physicist has a faculty for mathematics and is fascinated by its beauty and apparently unlimited possibilities. But he wishes also to be a physicist and to say something about the world of experience. What is more natural, then, than that, having a beautiful set of theorems panting for application to the physical world, he should exercise less than the necessary restraint in assuming a particular application which, if it were justified, would bring a large field of experience within the scope of his analysis? That this actually happens is not evident to the non-mathematician, but there are elementary examples from earlier times which have their close parallels in modern mathematical physics. Here is a mediaeval argument for immortality. If a equals b, then twice a equals twice b, whatever a and b may be. Now to be half dead is the same as to be half alive. Hence to be dead is to be alive. I think this helps us to understand why an unenlightened education in mathematics can have the effect of stultifying one's reasoning powers.

Bertrand Russell's collaborator in the great *Principia Mathematica*, the late A. N. Whitehead, shows, from another point of view, the same process in operation:

'By the aid of symbolism,' he wrote, 'we can make transitions in reasoning almost mechanically by the eye, which otherwise would call into play the higher faculties of the brain. It is a profoundly erroneous truism, repeated by all copy-books and by eminent people when they are making speeches, that we should cultivate the habit of thinking of what we are doing. The precise opposite is the case. Civilization advances by extending the number of important operations which we can perform without thinking about them. Operations of thought are like cavalry charges in a battle—they are strictly limited in number, they require fresh horses, and must only be made at decisive moments.'[1]

This was written some fifty or more years ago, when 'decisive moments', in the modern sense of the term, were rare, and civilization indeed advanced by performing operations without thinking about them. But we are no longer in that state. Almost every moment is decisive and calls for the clearest and deepest thought, and we are not prepared for it. To meet it we have an army of mathematicians who can indeed put one set of symbols into the analytical mill and grind out, with almost infallible accuracy, a consequential set, but they have lost the power of performing operations of thought. When presented with a piece of reasoning that requires such operations they become bewildered; they see only 'a smother of words'. Words will not go into the mill, so they can have no significance. This is not a nightmare; it is what is actually occurring. However incredible it may seem, we must recognize that if we are to bring reason to bear on our present situation, we must look for it elsewhere than to our mathematical physicists.

5

When it comes to the relation between mathematics and ordinary experience—not measurements which we make in our laboratories but that which happens to us naturally and constitutes the

[1] *Introduction to Mathematics*, p. 61 (Home University Library).

substance of our lives—the situation is even worse. Perhaps the principle of indeterminacy provides the aptest example of this, but that deserves separate consideration. I have said enough, I think, to enable me now to state my position with regard to your criticism of present-day mathematical physics, for, like mine, it arises from your perception of the gulf that exists between physical theory and experience. I wholly agree with you in this, and with you I believe not only that the gulf exists but that its existence is one of the most serious maladies of our time. But I am not able to share your view as to the proper treatment of the disease. If I understand you rightly, you would dismiss the abstract concepts of mathematical physics and substitute concepts more directly related to experience. Your energy, active or quiescent, is something of which we can form a mental picture; it is a constituent of a single objective world which is out there waiting to be discovered and is no less there whether we succeed in discovering and analyzing it or not. You would limit the activities of physicists to the examination of this unique objective world and so bring them back to reality.

That limitation I cannot accept. If we pursue this difference of view it will bring us back to the question we discussed at the beginning, namely, the meaning that is to be attached to the phrases, 'the universe as it is' and 'the universe as it appears to us'. We need not repeat that, but the form which our difference takes in this connection may perhaps best be indicated by extending the image I used just now. You would impose a three-mile limit on the adventures of the ship of mathematical physics beyond the shore of experience: I would leave her wanderings unrestricted but insist that she remains connected with the shore by a rope of indefinite length and infinite possibilities of sinuosity, which must never become broken or detached. We agree that at present she is distant and drifting. You would recall her; I would connect her.

Leaving metaphor, you would substitute for the abstract concepts of physics a concept of energy as an object of the same character as water (not material in the narrow sense, of course, but having the same kind of objective reality), and allow it to take different forms, just as water can exist as ice or steam. The operations of energy would then explain the various phenomena of nature, just as the transformations of water into clouds, rain, etc., explain certain meteorological phenomena. That seems to me to be a return to the pre-scientific type of philosophy whose success is incomparably smaller than that achieved by mathematical physics. The device of creating artificial experiences (measurements), expressing their relations by means of concepts whose only limitation is that their behaviour accords with the definitions freely bestowed on them, and applying the results to everyday experience in accordance with the correspondences that we find to exist, has proved so remarkably successful that I can see no reason at all for ceasing to employ it. I can find no grounds for rejecting the mathematical equations of physics or the concepts, however abstract, represented by the symbols in those equations (apart, of course, from possible technical details); my alarm comes from the indifference and, incredibly, the stupidity with which physicists regard the all-important question of the relation of their equations to experience. They have only to call a co-ordinate by the letter t, and they automatically think that a clock is bound to record the values that the equations give for t. That is indefensible, but it does not follow that t may not be an eminently respectable co-ordinate, having a relation to clock readings that may be put to good use.

LORD SAMUEL:

1

My comment on this must begin with a point that arises in your closing paragraphs where you give an outline of what you take to be my own general view on this matter, for that is not at all what I believe. You write (p. 148): 'If I understand you rightly, you would dismiss the abstract concepts of mathematical physics and substitute concepts more directly related to experience. Your energy, active or quiescent, is something of which we can form a mental picture; it is a constituent of a single objective world which is out there waiting to be discovered and is no less there whether we succeed in discovering and analyzing it or not.'

In the main, I accept what you assign to me as my concept of a given universe, real in its own right. But you go on to say that I would 'limit the activities of physicists to the examination of this unique objective world and so bring them back to reality'. That is not at all my idea, and I do not think that, when you come at the last stage to read again the script of this Dialogue as a whole, you will find any notion of that kind in any of my contributions.

On the contrary, let me refer you to a paragraph on pp. 43-4 in which I give a condensed summary of my thesis 'that there exists a universe, real in its own right and in no way dependent upon human perceptions or ideas; and that science and philosophy must study both that cosmos and also the man-centred world which is the outcome of those perceptions and ideas. Wherever the second is found to differ from its counterpart in the first, it is the business of science to try to bring them into conformity with one another. This can be done—has in fact continually been done, and is being done, piece by piece, more successfully today than at any previous time—along either of two lines of approach: either by making new discoveries about the cosmos and its processes; or else by further research within our own field of knowledge, resulting in fresh adjustments in our own ideas about nature: more often it is done along both lines simultaneously.' There is no suggestion here of any limitation of any kind on the activities of physicists. Having followed the wonderful achievements of modern science during the last nearly seventy years I should be purblind indeed if I were to undervalue the developments of mathematical physics during that brilliant period, or seek to discourage in any respect its speculations. Speculation has ever been the pioneer of discovery. So it has proved in full measure in our own day in this sphere, and so we may hope it will continue to prove, both in the immediate and in the more distant future.

But every inquirer will soon discover how many deficiencies remain. Most of the great scientific problems—some dating right back to the Greeks — have not yet been answered. Where mathematics has not been able to find a clue and comes to a dead end in some barren paradox, our present-day physicists often would have us, for that reason, refuse to attempt to find, or even to wish for, any answers. They would impose dogmatic pronouncements, of a metaphysical rather than a physical character, not supported by observation and experiment, nor yet by legitimate inference based on established experience. If we are to work towards a co-operation between philosophy and science, I fully agree with your claim that philosophers cannot be expected to acquiesce, humbly and tacitly, in pronouncements by physicists, however eminent, which may be the outcome only of translations of given facts into geometrical or arithmetical symbols, and the manipulation of those symbols in equations; ended by endeavours, sometimes perfunctory, towards re-translations from that world of metaphysical abstraction to the common world of fact.

2

In the course of these discussions you have consistently criticized such presumptions, and you have fortified your argument with definite examples. Especially you have dealt fully with one of the most topical of the present controversies — in which you are yourself the principal protagonist on the one side—that queer hypothesis that motion about the universe in spaceships at high velocity would decelerate material clocks and retard the ageing of human bodies. You also reject all theories based on the notion that there is such a thing as abstract time. As you recognize, we cannot, of course, dispense in practical life with particular 'times': I might say 'I must leave in ten minutes to catch the twelve o'clock train to Scotland, where I am going for a fortnight of my annual holiday'—our vocabulary would be very insufficient if it was not able to include sentences of that kind. And, when we wish to think with the aid of mathematics, there can be no objection to symbolizing such particular times as $t_1, t_2, t_3 \ldots$ But it does not follow that we are thereby entitled to take that long step further and imagine the abstraction of all such into a single universal time, and assume, without argument, that when we have designated it by a label 'T', we have conferred upon it reality. And if we analyze the myth of the space-travellers we shall at once discover that it is this fallacy of regarding time as 'something'—capable of affecting the course of physical events in our own sensory universe—that has misdirected this school of mathematical physicists.

3

But instances arise from time to time when ingenious minds are able to attract widespread attention among the lay public to other speculations of a similar order.

You may remember that a good deal of interest was aroused in the 1920s and '30s, by a theory put forward by J. W. Dunne, in a book with the title *An Experiment with Time*.[1] In order to

[1] Published by A. and C. Black, 1927.

furnish an explanation of alleged precognition in dreams of events which did afterwards happen, in exact detail, in real life, he offered the speculation that there might be in the universe several 'times'; or one 'time with several dimensions and a serial structure', so that the same event might happen twice over. This thesis he supported later in another elaborate book, with many diagrams and equations and a mystical background.[1] The idea greatly appealed to some of the amateurs in science and philosophy; among others to Mr J. B. Priestley, who wrote two plays based on it—*Time and the Conways* and *I Have Been Here Before*. They were, if I remember rightly, entertaining and successful plays. I had the pleasure of Air-Commodore Dunne's acquaintance, and, like all who knew him, respected the sincerity of his conviction that he had made a discovery of great importance to cosmology. I read the books, which he had been good enough to give me, but was not convinced. Something essential seemed to be missing. From the first volume page one was missing. *An Experiment with Time* ought to have begun with a statement why it should be believed that 'Time' existed, and be able, as an active agent, to perform any functions at all, including those that were there being assigned to it; and if so, through what physical mechanism. But Dunne did not attempt that. The word 'time' was used, without introduction or explanation, as though its existence as something were a natural assumption. To a reader like myself, who regarded particular 'times' as nothing more than measurements of intervals between events in the man-centred universe of our experience, and not as fractions of a universal abstract 'Time' which was an element in the given cosmos, the book could not get started without that missing page.

And now again we have this other strange fallacy—the myth of space-travel neutralizing 'time' and opening the possibility of man's escape from the infirmities of old age, or even, conceivably, from death itself. I would cite these as instances, among many that may be given, of the confusion that must follow from refusal to recognize the differentiation that should be made by both philosophers and scientists in the study of the two worlds.

But more important than either of those two debates has been the controversy about indeterminacy, arising out of Heisenberg's

[1] *The Serial Universe*, published by Faber and Faber, 1934.

principle of uncertainty, and leading to Born's assertion, which appears in his book *Natural Philosophy of Cause and Chance*[1] that 'Chance is a more fundamental conception than causality' (p. 47). This doctrine has lain heavily upon science, and therefore upon philosophy, ever since the revolutionary discoveries at the turn of the nineteenth and twentieth centuries.

4

I return now to the main subject of this Section—if science and philosophy are to work together, what part is to be played by mathematics? Many scientists hold nowadays that it should be the predominant part. This is the result of a combination of events: Michelson-Morley was the beginning; leading to the discarding of the hypothesis of an ether; and followed quickly by the discoveries of radioactive elements, of the electron, and of the structure of the atom—a nucleus and particles surrounding it. About the same time Einstein put forward his two theories of relativity, which came to be accepted by physicists generally. Next came Planck's quantum theory, developing into a whole new system of mechanics. The movements of atomic particles were found to be, in some respects, inaccessible to laboratory methods of research, because the application of those methods interfered with, and altered, the processes themselves that were being investigated; and this must forever be so. Hence Heisenberg's theory of indeterminism, or 'principle of uncertainty'. Finally, under the urgent stimulus of the needs of all countries for defence against the possibility of nuclear warfare, an intense and successful activity in nuclear research. All this was greatly assisted by the use of mathematical symbols and equations.

As a consequence, many of the leading minds in theoretical physics began to lose interest in what had always been regarded as the immediate purpose of science—the direct study of the phenomena of the universe and their causes; they turned to concentrate upon man's experience of those phenomena. The easiest line of approach was clearly through studying the inter-relation of one set of experiences with another, and this could best be achieved by their numerical and geometrical measurement. The furthest point was reached by Max Born, who postulated chance, and the laws of the probabilities of large numbers, to replace causality as the basic mechanism of nature.

[1] Oxford, Clarendon Press, 1949.

You have quoted Born in an earlier section of our Dialogue, but as this is the crucial point in what we are discussing now, I would like to recall another passage from his book *Natural Philosophy of Cause and Chance*: 'Classical physics had eventually to incorporate chance in its system. Today in physics, chance has become the primary notion, mechanics an expression of its quantitative laws, and the overwhelming evidence of causality with all its attributes in the realm of ordinary experience is satisfactorily explained by the statistical laws of large numbers.' He speaks of 'laws of chance', which 'certain observable events obey', and of 'the principle of molecular chaos'. He adds that this statistical interpretation 'is now generally accepted by physicists all over the world, with a few exceptions, among them a most remarkable one'—namely Einstein. And he speaks of 'the new statistical physics'.

But all this imposes upon mathematics a task far beyond its capacities. In the first place what is this 'chance', which is to be the foundation of the whole structure? We are constantly exhorted to put aside everything that is not amenable to observation and experiment: it would be very interesting to watch the working of 'the laws of chance' while 'certain observable events' are obeying them. In point of fact the very term 'laws of chance' is an obvious contradiction in terms—if there are laws there can be no chance, where there is chance there can be no laws. The only acceptable definition of chance is a merely negative one—Leslie Stephen's 'Chance is a name for our own ignorance'.

Let me give as an example the very temple of chance, a gambling casino. At the roulette table a horizontal wheel, divided into numbered compartments, is set revolving by the croupier, and he launches a ball into it in the opposite direction; when the wheel comes to rest and the ball drops into one of the compartments, that number wins. The results are, for us, pure chance, in the sense that there is no possibility of our predicting them. Yet, as a matter of fact, the difference of the results in each case depends upon two factors that are strictly causal—the velocity of the wheel when it is set revolving and the velocity of the ball when it is thrown in. And each of these depends upon the degree of effort on the part of the croupier in moving his arm. It would be possible to devise a machine which would start the roulette revolving and throw in the ball; and if the machine were nicely adjusted, and all interference by air-currents or otherwise excluded, the same number could be made to win every time. But the croupier's arm not being a machine, neither he nor anyone else can gauge the amount of force that is being exerted on any particular occasion. Once the movements of the arm have been made, the result will follow with certainty. The causes are there. The element we term chance consists merely in the impossibility of our ascertaining and measuring those causes.[1]

Nor is there any such thing as 'statistical laws' which can directly influence the course of actual events in the everyday world. There are rules of simple arithmetic which enable averages to be calculated of a series, for example, of annual figures relating to successive years, and such averages can be a guide in estimating, more or less approximately, the probability of similar figures applying to succeeding years if all the conditions remain the same. And such statistics are useful—indeed indispensable—in relation to disease rates and death rates, for the guidance of Public Health Authorities and of Insurance Companies, and it is true that thereby, indirectly, they may influence what will happen to individuals. But they have no place in a philosophical discussion about causality. If the Public Health statistics, for example, for a certain town of a million people, have shown for, say, ten years, a death-rate for a particular disease of round about 100, with an average of 105, and if in the next ensuing year it proves to be, say, 110—not even the most enthusiastic mathematician would claim that it was a 'statistical law' which caused the death of those persons. The causes were individual in each case, and are to be sought, not in the world of mathematics and prediction, but in the world of medical practice, hospital treatment, water supply, drainage, and the like, together with personal conduct and habits. A statistic is nothing but the tabulation of a number of units, and you can get out of the total nothing more than you have put into it in the units.

5

On this fundamental issue I would rather cite the opinions of qualified philosophers and scientists than ask the lay reader to attach value to mine. We have had in this century two eminent mathematicians who have been also eminent as philosophers— the co-authors of the *Principia Mathematica*, Alfred North

[1] This paragraph is repeated from my *Essay in Physics*, pp.27-8 (Blackwell, Oxford, 1951).

Whitehead and Bertrand Russell. You have already referred to them. Professor Whitehead says: 'Nature is a theatre for the inter-relations of activities. All things change, the activities and their inter-relations. To this new concept, the notion of space, with its passive, systematic, geometric relationship is entirely inappropriate. The fashionable notion that the new physics has reduced all physical laws to the statement of geometrical relations is quite ridiculous.' He says again: 'Mathematics is now being transformed into the intellectual analysis of types of pattern . . . The essential characterization of mathematics is the study of pattern in abstraction from the particulars which are patterned.' He speaks also 'of confining thought to purely formal relations which then masquerade as reality . . . Science relapses into the study of differential equations. The concrete world has slipped through the meshes of the scientific net.' He concludes, 'There can be no true physical science which looks first to mathematics for the provision of a conceptual model. Such a procedure is to repeat the errors of the logicians of the middle ages.'[1]

Lord Russell writes, in his recent book *My Philosophical Development*[2] (p. 277): 'Logic and Mathematics are the alphabet of the book of Nature, not the book itself'. He says also (p. 27): 'The entities that occur in mathematical physics are not part of the stuff of the world, but are constructions composed of events and taken as units for the convenience of the mathematician'.

As to 'the principle of uncertainty', I do not think that anyone, whether philosopher or scientist, has doubted that the situation is as Heisenberg has described it: namely that the movements of sub-atomic particles are often beyond the ability of human research to 'determine'—that is to say, to ascertain and predict. But the word 'determine' is itself ambiguous. Most physicists take it to mean also that, those particle movements being not *determinable* by scientists, must therefore not be 'determined', in the sense of 'caused', by the mechanism of nature; and consequently must be left to be governed by this factor named 'chance'. Chance itself can only be handled by the arithmetic of probabilities, and this, therefore, is taken to be the substitute for causality.

But Einstein does not think so. In letters to Born, of 1944 and 1947, Einstein wrote the sentence that has become famous, 'You

---

[1] *Modes of Thought*, pp. 25, 191; *Essays in Philosophy and Science*, pp. 83, 85.
[2] Allen & Unwin, 1959.

believe in the dice-playing god, and I in the perfect rule of law in a world of something objectively existing which I try to catch in a wildly speculative way . . . I see of course that the statistical interpretation . . . has a considerable content of truth. Yet I cannot seriously believe it because the theory is inconsistent with the principle that physics has to represent a reality in space and time without phantom action over distances . . . I am absolutely convinced that one will eventually arrive at a theory in which the objects connected by laws are not probabilities, but conceived facts, as one took for granted only a short time ago.'

Born, when quoting Einstein as though he were a unique exception to the general acceptance by physicists of the principle of uncertainty, has omitted other names of authority. Max Planck, in several short books dealing with his quantum theory, that have been translated into English, has dealt at length with this point, and dissociates himself uncompromisingly. I will quote these brief extracts from a number that might be cited. 'Today there are eminent physicists who under the compulsion of facts are inclined to sacrifice the principle of strict causality in the physical view of the world . . . So far as I can see, however, there is no ground for such a renunciation.'[1] Again, in a book entitled *Where is Science Going?*[2], with an Introduction by Albert Einstein, Planck asks 'Does science in its everyday investigations accept the principle of causation as an indispensable postulate?' and he sums up his argument in reply—'Some essential modification seems to be inevitable; but I firmly believe, in company with most physicists, that the quantum hypothesis will eventually find its exact expression in certain equations which will be a more exact formulation of the law of causality . . . Therefore it may be said here that physical science, together with astronomy and chemistry and mineralogy, are all based on the strict and universal validity of the principle of causality. In a word, this is the answer which physical science has to give to the question asked at the beginning of the present chapter.'

Lord Russell had summed up the whole matter when he wrote long ago about determinism[3]: 'In one sense of the word a quan-

---

[1] M. Planck—*The Universe in the Light of Modern Physics* (Allen & Unwin, 1931, p. 47).
[2] Allen & Unwin, 1933, pp. 141, 143, 147.
[3] Bertrand Russell, *The Scientific Outlook*, p. 109 (Allen & Unwin, 1931).

tity is determined when it is measured, in the other sense an event is determined when it is caused. The principle of indeterminacy has to do with measurement, not with causation.'

PROFESSOR DINGLE:

1

I am glad that you have elaborated your view of the function of mathematical physics, and also, paradoxically, that I gave occasion for you to do so, for the removal of possible misunderstandings is often the best means of making a point clear. What I had in mind in saying that you would limit the activities of physicists to the examination of the unique objective world was the impression given by such passages as that on p. 125: 'many—and apparently the great majority—of physicists seem to have come to the conclusion that the puzzle is insoluble, and prefer not to trouble themselves about it any more: they prefer to translate the situation into mathematical symbols, and to cope with it by means of differential equations, that have become ever more complicated and artificial'. This seemed to present the artificiality of mathematics as an alternative to the reality of the objective world, and to criticize physicists for choosing the former. But you have now explained your position in such a way as to remove misapprehensions of this kind, and I think our attitudes are not so different as my earlier remarks might have seemed to imply.

2

Your reversion to the question of time from a new angle—that of the distinction between the mathematical symbol called time in the physical equations, and the supposed objective element of the universe which is called by the same name—raises questions which we have not yet examined explicitly. It is, I believe, the failure of physicists to understand the difference between *an objective physical event*, like the position of the hands of a clock on a particular occasion, and *the time at which the clock-hands have that position*, that has given rise to the illusion of asymmetrical ageing. The time of an event is a mathematical concept, which, according to Einstein's theory, can have as many values as we care to give it: the event itself is quite outside our control. The clock-reading at an event is definitely such-and-such: if we choose a time for the event earlier than that denoted by the clock, then the clock is fast; if we choose a later time, then the clock is slow. The same clock can be simultaneously correct, fast and slow, according to the time-system we care to choose; but the clock, and our physical bodies, like Gallio, care for none of these things.

Unfortunately, however, these two quite distinct things—physical events, like clock-readings, and arbitrary concepts, like times—have become so confused that when one has to deal with any passage in which the word 'time' occurs, it is often extremely difficult to sort out the various implications and to estimate the passage at its true significance. Dunne's ideas afford an excellent example of one facet of this problem. There are many others, but this one is worth considering for its own sake. I also knew Dunne, and respected not only his sincerity but also the courage with which he tackled problems into which his theory led him but for which he had not had the necessary training. I first met him when he was in course of writing *The Serial Universe*, and read and discussed with him the MS. of that work. This, of course, was after the publication of *An Experiment with Time*, and the contrast between the main themes of the two books illustrates very well the point I now wish to make.

An Experiment with Time, as its title indicates, is mainly concerned with experience and with events that are directly experienced: *The Serial Universe* is concerned with conceptions—chiefly the conception of time, which in this book has ceased to be directly related to experience and has become a concept of the kind that mathematicians represent by symbols. The two themes are quite distinct. It is possible that the precognition supposed to be afforded by dreams may be established and the theory of the serial universe discarded, or that the precognition may be illusory and the theory adopted; or, of course, both may be accepted or rejected together. Let us take the dreams first.

Whether or not one can, in a dream, have foreknowledge of an event, is entirely a matter for experiment to determine. It does not matter at all whether you believe that there is such a thing as time in the universe or not; you may certainly discover, after dreaming of something, whether that something actually occurs or not. Consequently I do not miss your page one in *An Experiment with Time*: whatever might have been written on it could make no difference to the assessment of the evidence later advanced for precognition. Dunne recognized that his evidence, though very suggestive, was inadequate, and proposed further experiments. Unfortunately his interest later shifted to his theory, and he did not pursue this side of the investigation.

The great difficulty, of course, in this purely experimental problem, is to decide what degree of resemblance between dream and later happening is sufficient to establish more than an accidental relation between them. If we ask for *complete* resemblance then it is practically certain that this has never occurred, and absolutely certain that it is not a normal phenomenon. On the other hand, if we are satisfied with only the most general resemblance, then all dreams are prophetic: I dream that I am awake, and it has always come true. Between these extremes there are all shades of possibility. You dream you are in the company of other persons, and it usually happens: you dream you are in the company of a particular person, and it sometimes happens: you dream you are at some particular place, talking to a particular person on a particular subject, and it rarely happens. How can you apply the theory of probability to decide what degree of repetition justifies the assertion that an objective relation exists between dream and occurrence?

Dunne ingeniously surmounted the objection—fatal to most theories of dream precognition—that we could prevent the fulfilment of the dream by avoiding the circumstances in which the event is to occur. You did not, in his view, dream of each event as a whole, but in parts which, in the re-enactment, were separately associated with parts of other events. Thus, if you dream first that you are knocked down by a car in Regent Street, and later that you are shopping in Oxford Street, you may shop in Regent Street and be knocked down in Oxford Street. The details of each partial event may be so faithfully reproduced as to exclude chance, while the association of the parts might be quite haphazard.

I do not think there is as yet sufficient evidence for this supposed phenomenon to justify belief in its actuality, but I have no doubt that it must stand or fall by the evidence of experience. Attempts to dispose of it on the grounds that precognition is impossible are to my mind quite misguided. They rest, in effect, on a tacit supposition that there is an objective something called time, which has certain properties that prohibit the experience in question. But it is quite imaginable that we might obtain experiences for which such precognition is the most reasonable explanation, and, that being so, this prohibiting time must be a phantom. I do not think Dunne needed to insert a page one to introduce it, since it could do nothing but impose a false veto on the experiment which he was quite justifiably proposing.

But it is true that *An Experiment with Time* included an adumbration of the idea of 'serial time' which he was later to develop in *The Serial Universe,* and I agree with you that that implied the existence of an objective time which should have been established before its nature was discussed. The 'time' which he was analyzing was, in fact, a concept—i.e. something which we deliberately introduce as part of our mental apparatus for organizing experience into a rational system—and as such it is something which we are at liberty to define in whatever way we find most suitable for the purpose; it is not something that exists outside us, which we must accept and describe as it is. Dunne was partly aware of this, and I found his attitude curiously inconsistent in that, while he recognized the arbitrary character of the time concept, he nevertheless treated time as though it had an objective nature which we might discover but not alter.

Comparing his view of serial time—i.e. an infinite succession of times—with the ordinary view of a single time, he likened the views to two different expressions of the fraction 'one-third'; we could represent this by $\frac{1}{3}$ or by $0 \cdot 33333 \ldots$ In the first case we had a single symbol, in the second case an infinite sequence of symbols. Each of the 3s in the second representation corresponded to one of his 'times', but you could take them all together if you wished, and speak of a single time represented by $\frac{1}{3}$. It was a good illustration, showing clearly the purely conceptual character of 'time'. But then he quite surprisingly converted each 'time' into a separate scope for experience, and held that while you might die in time 1 you could still go on living in time 2—as though the first and second 3 in the recurring decimal each had an independent significance of its own. Here it was he, and not his critics, who transgressed by introducing a quite unwarranted objective 'time' into philosophy.

In so doing I think he may have been his own worst enemy. The conception of serial time is a perfectly legitimate one, and might be an extremely useful one. Some of Dunne's comments on physical problems, in fact, suggest that, in the hands of a

L

competent mathematical physicist, this conception might have led to valuable results. But this possibility was discounted in advance by the false notion that 'time' either was or was not serial, and if you decided that it was not, then 'serial time' deserved no further notice. It is an excellent example of the confusion of what you would call the objective world, and I would call experience, with what we would both call a mathematical concept.

3

But now let us come to your main point in this section, which I take to be this. Physicists have discovered in mathematics an excellent tool for investigating the universe, and it has already helped them to the solution of a number of problems. But others remain unsolved, and physicists, instead of persevering or trying some other method, have elevated their tool to an end in itself: they have substituted for the real problems pseudo problems that originate only in mathematical conceptions, and having solved these to their satisfaction they consider that the real problems no longer exist. Conceptions of their own devising, like 'time', 'chance', and so on, are represented by symbols—t, $P(p/q)$, and such things; equations are formed in which these symbols occur; they are found to hold together very neatly and, hey presto, we now know all about it. We need not ask what keeps a traveller young; t can be made to change very slowly, so that settles it. It is childish to inquire, like Newton, into the causes of things; $P(p/q)$ has conquered causes. The universe is a mathematical symbol—or, to express it in terms understood of the people, a thought in the mind of a Great Mathematician. When you have found the relations between the symbols you have understood the universe.

To this you object, and I agree wholeheartedly: it is inexcusable. But the ground of my objection is not identical with yours because as I have already explained, I cannot bring in the 'universe' which to you is essential. Its equivalent in my view is what you would call our experience of this universe—or rather the conception which that experience leads us to form but which has no other ground for claiming to 'exist'. Experience is to me our *primary* datum, simply because it seems to me obvious that it is. Even if there is a universe behind it, it is only through our experience that we can know anything about it, and therefore our philosophy, if it is to be defensible, must in the last resort be based on that and must justify itself by appeal to that.

Pure mathematics, in itself, is independent of the universe and of our experience of it: it is an expression of rational necessities that hold good quite apart from what our senses present to us. When we agree to the proposition that things which are equal to the same thing are equal to one another, we make no pronouncement about what the 'things' may be, nor do we even imply that there are, in fact, any 'things' at all, but if there are, then this is true of them. All mathematics is ultimately of that character. Our object in science or philosophy is to find rational relations between our experiences (or, as I expect you would prefer to say, between those elements of the universe that cause them), and when we detect such relations we express them in mathematical form because mathematics is the language of pure reason.

For example, if we find that one grain of sand is balanced by a certain weight in a pair of scales, and that a second grain is balanced by the same weight, then we also find as a matter of experience, that the first grain is balanced by the second. We therefore feel justified in inventing a 'property' of a grain which we call its *mass*, and in applying to mass that branch of mathematics that includes such axioms as that things that are equal to the same thing are equal to one another. But it is perfectly conceivable that each of the grains might have balanced the same weight and then have been found *not* to balance one another. In that case we should not have found this concept of mass of any use, and it would not have been introduced into physics in this connection. We often see a striking likeness between a child and its father, and also between the child and its mother, whereas the father and mother are not at all alike. The concept of 'likeness' in this sense accordingly finds no place in science. (Incidentally, in the mathematics of infinity, there are things that are equal to the same thing but not to one another: mathematicians are well provided for emergencies in physical investigation.)

The relations between experience and mathematics are entirely of this character. Mathematics is a system of relations between purely ideal concepts. We can form such concepts *ad lib.*, defining them as we wish, and then work out an elaborate system of relationships that follow by pure reason from our definitions. All this, as I explained in an earlier passage, is quite independent of the world of experience. We look at our experience and select those elements of it that we can consistently represent by the concepts of some mathematical system, and then we can say that the experiences are related in the same way as the concepts in that system. We know of no *necessity* that this should be so. It is no more necessary that two elements of experience which are equal to the same thing shall be equal to one another that that two elements of experience which are like the same thing shall be like one another. We just find that such-and-such facts of experience can be consistently correlated with such-and-such mathematical concepts, and when that is established the whole field of relations between those concepts, which has been worked out mathematically, becomes applicable to the whole field of such facts of experience. Every now and then we find that we were mistaken in our correlations. For instance, the concepts of mass, force, etc. which are united in an elaborate system of Newtonian gravitational theory, were long thought to correspond exactly with the experiences we get when we apply our measuring instruments to the bodies in the solar system, but at last the correspondence broke down. That particular mathematical system has therefore been discarded as a true representation of the relations between the experiences, and another, containing quite different concepts (such as curvature of a space-time field, for instance) has been adopted in its place.

Now this has not been at all understood. The mass of a body, for instance, has been thought in the past to be an element of experience—or, in your terms, as a property of a body in the objective universe—and so to be something *given* us, which we have no *right* to discard. We can now see that that was an error. It was not given us; we invented it because it enabled us to use a particular mathematical system. Now that that system has failed us we have no need to invent it, so we simply let it die. (For convenience, of course, we still use it in limited problems, but we do not regard it as having any fundamental significance).

This discovery of the true relation between mathematical concepts and the facts of experience is of fundamental importance. I think, as I have said so many times already, that it makes it impossible to maintain the view that the quantities that we talk about in physics—such as the length and mass of a body, the heat content of a system, the electric potential at a point, and so on—which are represented by symbols in our equations, stand for properties of an objective universe, existing in its own right, which we may try to discover but cannot create or destroy. They are freely chosen concepts whose importance arises entirely from the fact that their relations with one another are paralleled by the relations between our experiences. The moment they cease to exhibit that parallelism they dissolve into thin air.

Mathematical physicists have to some extent brought their (usually unconsciously held) philosophy into line with this, and to that extent I find myself in sympathy with their refusal to accept your criticisms. (By mathematical physicists I mean what may be called *orthodox* mathematical physicists — those who make such statements as those to which you are here objecting. I am aware, of course, that there are a minority who are uneasy at this.) Some of the questions which you say they have evaded, in so far as they depend on the previous acknowledgment of a universe outside experience, existing in its own right, are, I think, not questions to which it is possible to give answers because they are framed in terms which we can now see to be incompatible with our possibilities. But those questions can be framed also in terms of experience, and my criticism of the predominant philosophy of mathematical physics is that they have not been clearly framed in those terms and so have not been faced. Physicists still retain the fiction that their symbols represent properties of an objective universe, and so mistake ideas, which they are quite justified in forming as tools, for discoveries about nature. The point is subtle, but I will explain what I mean as clearly as I can.

Let us take any physical equation at all which is recognized as being important: what actually does it mean? It consists of a number of symbols arranged in two sets, of which one set is said to equal the other. For example, take the equation $F = G\dfrac{Mm}{r^2}$ which is Newton's law of gravitation. Here each letter (except G, which is merely a number whose value can be changed when we change the units of measurement so that the equation can hold good for all possible units) stands for a number, and that number is ideally the reading of a particular kind of measuring instrument when applied in an understood way to some object. In this particular equation two objects are involved—two bodies situated at different points in space. M is the mass of one body, m that of the other, r is the distance between them, and F is the force of attraction between them. Now each of the symbols, M, m, r and

F, stands for the result of a particular operation of measurement. If, for instance, as in a famous experiment made by Cavendish, the bodies are two lead balls, M and m are obtained by weighing the bodies in an ordinary balance; r is the distance between them, measured by a standard metre scale; and F is the force we have to exert so as just to keep the balls from moving towards one another—again measured by a standard process. This is a type of all physical equations—though, as we shall see, it is a particularly simple one. In general, every such equation, when written so as to express explicitly what actual facts of experience it relates together, is of just this character.

Now, strictly speaking, this is all that we can say about the matter with which the equation deals: it expresses the fact that the results of a number of measurements, each in itself independent of the others, are nevertheless related in this particular way. Physics consist at bottom of a collection of such relations, and, nothing else. But we draw a picture around this nucleus, in which the symbols are transformed from the results of performing operations on bodies into properties assumed to be possessed by the bodies themselves. M and m are properties (*their* masses) of the lead balls, and r we regard as the extent of a certain portion of space; but we do not usually think of its extent as a property of space in quite the same way as we think of its mass as a property of a ball. F is a sort of half-way case. It has often been regarded as the stress in an ether existing in the space between the balls, and so a property of the ether in much the same way as its mass is a property of a ball, but this interpretation has not been universally adopted, although those who favour it have sometimes regarded it as a necessary interpretation.

But the important point to notice is that this picture, indispensable as it may seem (and, humanly speaking, probably is) for the purpose of making progress, is not *necessarily* involved in the situation. It is contributed *by us*, and can be withdrawn by us without altering anything in the actual relations between our experiences, as I should say, or in our discoveries about the universe, as you would say. Despite the different mental pictures we form of the quantities F, M, m and r, those quantities are all fundamentally the results we obtain by performing specified operations, and so of essentially the same character. What the equation inescapably says is only that the results of performing these particular operations are related in this way, and it is important because it holds good no matter how freely we change the bodies on which the operations are performed or wherever we place them.

Now let us consider another equation, $\frac{Ee}{a^2} = \mu\omega^2 a$, and to remove possible perplexity I will say at once that it is what is usually described as the (simplified) equation of motion of an electron around a proton in the hydrogen atom. In the picture which we form of this, E is the charge of the proton, e that of the electron, a is the radius of the orbit, μ is the mass of the electron, and ω is the angular velocity of its revolution. In appearance this equation has the same general form as the other, and we should therefore expect the same sort of relation to exist between the picture and the bare essence of the situation round which the picture is drawn. We should expect, for instance, that μ is the result of applying the operation of weighing to the electron, a that of measuring the distance between the proton and electron by a metre scale, and so on. But in fact nothing of the kind is true. Not one single quantity in this equation is the result of applying any operation, even remotely resembling that which the name given to the quantity suggests, to a proton or electron or anything at all.

Let us take one of them as an example—the quantity μ, say, which we call the mass of the electron. If μ is not the result of weighing an electron in a balance, what is it? To discover that we must look at the process by which its value is obtained, and we then find that that process consists of a number of complex operations in which a number of readings of measuring instruments are taken and combined together in a particular way. It would be far too big a task to describe this completely here, so I will substitute a simpler imaginary version in which we make a number of measurements, P, Q, R, S, T, form the quantity $\frac{P Q^3 R^{\frac{1}{2}}}{T^2} + S^5$, and then denote this quantity by the symbol μ and call it the 'mass of the electron'.

Now each of these symbols, P, Q, R, S, T, arises entirely from ordinary large-scale operations with visible and tangible bodies; 'fundamental particles' do not enter into the process at all. μ is *completely* defined in terms of gross matter, and the 'electron', the entity whose mass it is supposed to represent, belongs *only* to the picture which we voluntarily draw around the actual inescapable facts. Exactly the same thing applies to every other symbol in the equation, and so, if we were to write this equation so that it directly stated the facts it represents, we should get a most unwieldy expression, far too cumbersome for practical use, but one in which each symbol meant what it said, so to speak, a 'mass' being the result of an actual operation with a balance, a 'length' the result of an actual operation with a measuring rod, and so on.

But now, why do we select, out of this very extended and involved combination of symbols, the particular group, $\frac{P Q^3 R^{\frac{1}{2}}}{T^2} + S^5$, and call it μ, as though it had some sort of separate existence? In the first place, because we find that this group also occurs in other equations, where it has the same value as here. This allows us to take it as a whole and give it a name, and the name we give it is the 'mass of the electron'. But this is immediately to introduce our voluntary picture. We suppose the presence of an object, which we do not observe, endow it with a property called mass, and suppose μ to be the result we should get if we could put this imaginary particle in a balance and weigh it. This goes far beyond anything which the repeated appearance of this group of symbols justifies us in asserting. That permits us to represent it by a single symbol and give it a single name, but not a name that implies the existence of a particle possessing a property. If we are determined to choose a phrase instead of a single word we should write it 'mass-of-the-electron', and then we should not be able to take the illegitimate step of supposing that there is a particle called the electron hidden in the apparatus, which would be there whether we found its mass or not, and we should be kept to the strict truth that we had identified a combination of measurements which (for reasons which I do not want to complicate the matter by explaining, but which are quite independent of any hypothesis or pictures) we could properly regard as a 'mass', i.e. as equivalent to the result of performing a single operation of weighing.

But there is, of course, a reason why we speak of the 'mass of the electron' instead of the 'mass-of-the-electron', and it is this. Exactly the same process as that which I have just described leads to the isolation of another group of symbols in our completely expressed equation, which we denote by e and call the 'charge of the electron'—though, so far as our considerations have yet gone, we ought to call it the 'charge-of-the-electron'. This also occurs as a whole in various relations, and we have the same reason for calling it a charge (i.e. the result which we should get if we applied the standard process for measuring electric charges) as we have for calling μ a mass. Now the additional fact to be noticed is this. We have never found that e occurs in any situation in which μ is absent, or *vice versa*, so that *if* we adopt a picture in which μ stands for the mass of the electron, we must give that electron a charge e. Going further, we find that similarly there is a group of symbols—each again the result of a large-scale measurement—which we can picture as a rotation, and we call it the 'spin of the electron'; this also invariably occurs in situations in which μ and e appear. Hence, once grant the legitimacy of drawing imaginary pictures round the bare facts of observation (and for *practical* purposes no one would question it), and you are led almost inevitably to the picture of a tiny particle, with an extremely small mass and charge and a definite spin; and if, further, you suppose that what you are doing in physics is examining the structure and properties of an external universe, then there is virtually no alternative to supposing that this particle is indeed a real constituent of that universe, which, though it is inapprehensible directly by the senses, you have discovered by your researches.

But now comes a difficulty. If we suppose that the electron is 'there', in the same sense as the lead balls are 'there' in the Cavendish experiment, then it must occupy some definite position in relation to the visible part of the universe. Also, since, as it is necessary to suppose, it is frequently moving and has a mass, it must also have a definite momentum. There should therefore be groups of measurements standing for its position and its momentum. But in fact there are not. In a situation in which we can definitely say, 'this group must represent the "position of the electron" ', we find that the group which must then logically represent the 'momentum of the electron' is *necessarily* not present, and *vice versa*. It is as though the lead ball were quite definitely moving through the air with a precisely ascertained speed, but was nevertheless nowhere during the process. This is quite unimaginable.

And there is worse to come. In cases where the position group does appear, it is quite impossible to say how many electrons there are because it is impossible to count them. This is not because of the practical difficulties but because they are *by their nature not countable*. This we must conclude for the following reason. To explain the facts of observation in a situation in which, according to the picture, large numbers of electrons are moving about, we must suppose that when any two of them change places, nothing at all has happened. It is not sufficient to say that the second configuration is *indistinguishable* from the first: it must be the *same* configuration, or else the statistical results that follow from the proposed situation do not agree with observation. This is inconceivable with ordinary objects. If two people change places, then, even if they are identical twins, something has happened that is potentially detectable. But with electrons nothing has happened. We must certainly say that they have changed places, and we must also certainly say that no change has occurred. Unless we admit this the whole conception of particles falls to the ground.

The position is therefore this. We find a large number of relations between measurements which we actually make, showing that our experiences are not unconnected with one another but form a closely related system. We can express some of these relations very simply by imagining a picture in which unobserved particles, of the same nature as observed bodies, go through certain movements, but when we try to extend this picture over other relations we fail unless we endow the particles with properties quite inconceivable in ordinary observable bodies. Are we, then, to suppose that these particles 'really exist' or not?

Born, with his usual faculty for hitting the nail on the head (but, unfortunately, without a comparable faculty for perceiving where he is driving it) puts the problem very clearly in the book to which you have referred. He realises the dilemma, and comes down firmly on the side of the 'reality' of the electron. On p. 105 he says: 'Though an electron does not behave like a grain of sand in every respect, it has enough invariant properties to be regarded as just as real'. But then, why may not one equally well say: 'Though an electron behaves like a grain of sand in some respects, it lacks too many invariant properties to be regarded as just as real'? The resemblances and differences are comparable in number and importance — if anything, the differences predominate. There is nothing at all in the facts we know to make the choice anything but a matter of taste.

But surely it is quite fantastic to suppose that the matter is to be decided by voting. The electron is of quite a different nature from the grain of sand *because it enters physics in a different way.* We do not decide whether Hamlet really lived by comparing his behaviour with that of a living person. We know that our knowledge of him comes from a quite different source from our knowledge of, say, the actor who portrays him, and we have no hesitation whatever in saying that Hamlet has not the same kind of 'reality' as the actor. For exactly the same reason we must say that the electron has not the same kind of 'reality' as the grain of sand. The former 'exists' because we invent it to give a picturability to the facts expressed by certain equations, and the latter presents itself directly to our senses: the mass of the former is a shorthand sign for a combination of measurements, while that of the latter is the result of an operation of weighing; and so on. Yet we gravely discuss the 'existence' of the electron as though the question were quite independent of these differences.

The matter is carried almost to the stage of farce in the detailed examination which is made of the process of finding the 'position' of an electron. You imagine yourself making the particle visible by shedding a beam of light on it, whereupon it is knocked out of its position so that you cannot tell where it was (this is rather simplified but not distorted). But there is no such thing as the position of an electron: there is the 'position-of-an-electron', and to find that you would have to make measurements with clocks and scales and galvanometers and all sorts of things on complicated large-scale systems and combine the results in a certain way. It is laborious, and needs skill, but the process is quite specific. It is altogether a different process from that of measuring the position of a star in the sky. In that case the first thing to do is to identify your star, and that is just what it is impossible to do with the electron because there is no such thing. There are the 'mass-of-an-electron', the 'charge-of-an-electron', but no 'electron': no symbol or group of symbols in any equation stands for an 'electron'. Consequently this imaginary measurement of the position of an electron is sheer fantasy, and the extent to which it has bewitched physicists is a measure of the price we pay for neglecting to study the philosophy of science.

Even that falls short of the ultimate absurdity, because, if this 'proof' of the impossibility of finding the position of the electron were valid, it would discredit the very conception which it is supposed to support. It would prove the impossibility of determining the simultaneous position and momentum of the electron, but only by assuming that the electron *had* a simultaneous position and momentum: without that assumption you have no right to say that the light would meet it and knock it out of position; such statements would be meaningless. But it is essential to the theory, not that the electron has these properties of a material body which happen to be undiscoverable, but rather that its nature is such that these properties have no significance with respect to it. So what possible sense can there be in proving that if a thing existed you could not find it, when what you want to prove is that it doesn't exist?

And now I think I can set out very simply my view of this question of indeterminacy. The facts are that we have discovered a large number of relations between the results of definite processes of measurement. These are, of course, subject to the uncertainty arising from our natural liability to error and the necessary imperfection of our instruments, but we have no evidence in experience for the supposition that such uncertainty is not reducible without finite limit. We have therefore no reason to doubt (though, of course, no reason also dogmatically to assert) that *all* our experience—in physics at any rate—is capable of being related together in a determinate way. On the other hand, we have not so far been able to construct an imaginary picture of a hypothetical 'world' which we can regard as causing that experience by its impact on our sense organs, without peopling that world with entities whose individual behaviour cannot be determined but whose statistical behaviour is determined with so high a degree of probability that we have never observed any variation in it.

I see no reason for trying to go beyond these facts. But the orthodox mathematical physicist does go beyond them, and supposes that the imagined undetermined entities are, in fact, actual constituents of a 'universe' which exists prior to our experience of it. He accordingly concludes that that universe is at bottom undetermined, although most of our experience, since it is associated only with the behaviour of very large numbers of the undetermined particles, appears as though it were determined.

You, I believe, would agree with the mathematical physicist in his postulation of the 'universe', and I expect that you would be willing to accept the establishment of fundamental particles, such as electrons, as elements of that universe. But you would not admit his right to regard his ignorance of the behaviour of the particles as an establishment of the actual irresponsibility of their behaviour.

If the assumption which both you and the mathematical physicist make is justified, I have no hesitation in taking your side. We set out to discover the laws of the universe. We find that the universe is made up of particles, the laws of whose behaviour elude us. We do, however, manage to show that the regularities in our experience can be accounted for without a knowledge of such laws. The mathematical physicist thereupon declares that there are none. But since all the discoveries of physics — the relations between our experience — would be exactly the same whether there were such laws or not, this conclusion is quite invalid: the problem with which we began still remains. This is your contention, and I see no answer to it. So long as the task of science is regarded as the discovery of the laws of the universe, then it is merely shirking that task to say that there are no such laws, for that cannot possibly be deduced from the fact that we don't need them for some other task.

But to me the task of science is not the discovery of the laws of the universe, but that of the regularities in experience. I have no objection to particles whose behaviour is due to chance because the particles are merely tools of our own devising, and if they fulfil their purpose without having to possess a known code of individual behaviour, I am perfectly satisfied. Shakespeare tells me vividly what sort of man King Lear was by saying that he was every inch a king. I am not perplexed by not knowing how many inches he had, and I do not care in the least whether that number was ascertainable or whether the act of measuring him would have made him quiver so much that it became indefinite. It is the character of Lear that Shakespeare has to depict, not the details of the literary machinery that he employs for that purpose.

Born, again, in metaphorical fashion, admirably refutes his own view in his reply to Einstein. Commenting on Einstein's disbelief in the dice-playing God, which you have cited, he says in effect that, whether or not God throws dice, we have to throw them in order to discover what he does. Exactly: the dice-throwing is all ours.

LORD SAMUEL:

There is only one comment that I wish to make on your last piece, but it is on a matter of importance. You criticize the mathematical physicists for their attitude to the unsolved problems in the behaviour of the sub-atomic particles, and you approve my unwillingness to accept a 'principle of uncertainty' as a means of by-passing those problems. So far, so good. But your own argument is also mathematical, and it involves a postulate of wider application to which I cannot subscribe. You agree that the universe consists of particles, but you say that those particles are 'tools of our own devising' (p. 173); and if 'the laws of their behaviour elude us' all we have to do is to devise different tools better adapted for their purpose. You say that to get tangled up with the question whether electrons 'really exist' is a mistake; it is not likely to lead to a definite result, and therefore cannot serve any useful purpose.

I am glad that we are at one as to Heisenberg's principle of uncertainty; but cannot concur in your further conclusion. I share the general belief that the existence of the particles is a reality; that the vast nuclear technology that has been developed as a consequence of that belief has rendered important practical results, for good as well as (we may hope only temporarily) for evil; and that it is impossible to put it aside until some better mathematics has been devised to meet the present situation.

In an earlier section in which we gave a preliminary discussion to this point, I offered two reasons for believing in the reality of the electron, the neutron and the rest.

The first rested upon the fact that radio-active particles are known, from tragic experience, to be injurious, and even fatal to human life. So much so that the authorities in all the countries where atomic energy establishments have been set up, have been obliged to isolate the workers in the departments where streams of those electrons, neutrons and others might possibly escape, and protect them by thick walls of concrete faced with lead. I instanced also the Japanese fishermen in the Pacific who were struck by a fall-out from the explosion of an American atom-bomb; some were killed and some infected with cancer of the blood-cells (leukemia). So clearly was this established, that the American Government recognized its liability, and paid a considerable sum in compensation. And I ended by saying that a differential equation cannot give a fisherman cancer.

You do not question the accuracy of these facts, but you say that you could imagine a witch-doctor in some primitive tribe frightening some credulous victim of his incantations so as to cause his death. That also is true, but is it relevant? Suppose that the Americans had refused to pay compensation; that the Japanese Government had brought the case before some international tribunal; and that there the defence had offered incantations, or any kind of nervous shock, as the cause of the injuries and as a reason for non-payment, adding that scientists feel no certainty that electronic particles have any existence as physical facts. Would not the applicant's counsel at once have asked whether the American Government had not itself spent many millions of dollars in protecting the staffs employed in their laboratories where radio-active particles might escape, and similarly the crews of their nuclear-engined submarines; and whether this did not prove that they themselves did not believe those particles to be merely mathematical symbols, but were physically real, 'as real as grains of sand': would not the inevitable answer in the affirmative bring the case to an immediate close, with substantial costs added to the compensation?

My second argument rested upon the universal acceptance of the principle of the Thomson-Rutherford theory of the atom, namely that all chemical atoms have a structure, and that that structure consists of particles—that is of discrete entities of some kind—in rapid motion within the confines of the atom. I gather that you do not reject that theory and would not hold that it ought to be dropped from the teachings of modern physics. Nor do you dispute that physical objects, like stones, or our own bodies, do exist. Your quarrel is only with such alleged characteristics of the particles as mass, momentum, position, spin, and the like—and perhaps also reactions with wave-patterns. But the failure of physicists to find the right mathematics for dealing with those riddles surely cannot affect the question whether particles exist or not.

If they do not exist, what becomes of the atom? On the principles now accepted we have the series, passing from macrocosmic to microcosmic—material objects, consisting of molecules, which consist of atoms, which themselves consist of particles. If the particles are eliminated from the series as being only mathematical, what becomes of all the rest? If you deny reality to the particle you deny it to the atom also. If the atom is dematerialized, the molecule, composed of atoms, must go with it: consequently all physical objects, including the organic cell and all living organisms, must go too. With a bit of mathematical logic you will have done something far more destructive than a thousand hydrogen bombs; you will have dissolved the whole universe into nothingness.

You have brought in Prospero in an earlier context: I would invoke him again, in his most familiar lines[1]:

> Our revels now are ended. These our actors,
> As I foretold you, were all spirits, and
> Are melted into air, into thin air:
> And, like the baseless fabric of this vision,
> The cloud-capp'd towers, the gorgeous palaces,
> The solemn temples, the great globe itself,
> Yea, all which it inherit, shall dissolve,
> And, like this insubstantial pageant faded,
> Leave not a rack behind.

I am sure you would not have intended to cause so much inconvenience, but this is what you will inevitably have done if you have succeeded in disestablishing the electron. We see now what Whitehead meant when he said that 'Science relapses into the study of differential equations. The concrete world has slipped through the meshes of the scientific net.'

So at the end we shall find ourselves obliged to set mathematics gently aside for the time being, and go back to the obvious, to common-sense and to Dr Johnson kicking the stone.[2] Then you will be able to march away triumphant, with your Human Experience—but I too, with my Objective Reality.

---

[1] Shakespeare—*The Tempest*, Act IV, Sc. 1.

[2] *The Life of Samuel Johnson* by James Boswell. August 6th, 1763.
'After we came out of the church, we stood talking for some time together of Bishop Berkeley's ingenious sophistry to prove the non-existence of matter, and that every thing in the universe is merely ideal. I observed, that though we are satisfied his doctrine is not true, it is impossible to refute it. I never shall forget the alacrity with which Johnson answered, striking his foot with mighty force against a large stone, till he rebounded from it,—"I refute it thus".'

PROFESSOR DINGLE:

I think we must agree to differ on the question of the 'existence' of fundamental particles, and I would only add a brief note on the two reasons which you give for your view on this point.

The Japanese Government's claim, in my view, could be properly based only on the fact that the American Government had performed certain large-scale operations which it knew would result in injury to Japanese citizens. The claim would stand or fall on consideration of this fact, no matter what might be the physical explanation of the occurrence. Indeed, if the American Government tried to put the blame on sub-atomic particles and so to exculpate itself, the tribunal would have little difficulty in passing judgment.

There is indeed a sort of continuity from large-scale objects down to the molecule, the atom, the electron, but continuity does not imply identity. If you look at a straight rod, made of some ductile material, by ordinary sunlight, you see simply a straight rod. If you stretch it, so that it gradually becomes thinner, then for a time you still see a straight rod, though you may have to resort to a microscope for the purpose, but a point is reached when it appears to widen, although you are still stretching it, and soon you see, instead of a single rod, a series of parallel alternately dark and light rods. The change is continuous, but what you see when you look at a very thin rod is quite unlike what you see when you look at a thick one.

I do not think you would hold that the series of rods is just as 'real' a description of the object as is a single rod. I need not here analyze the transition, but I do not think I am more vulnerable to the charge of dissolving the universe into nothingness than you would be if you declared the parallel dark and light bands to be an optical illusion.

IV

LIFE AND MIND

LORD SAMUEL:

1

Throughout the previous sections of this Dialogue we have been careful to make it clear that we have been dealing only with the material phenomena of nature, and have reserved the problems of life and mind. But these, even more than the others, demand that close co-operation between science and philosophy for which we wish to plead. And if we were to attempt any comparative scale of values we should probably agree that, for the human race, life and mind are more important than matter. Yet, now that we have arrived at that stage of our discussion, for my own part—I do not know how you feel about it—I find that there is much less that I wish to say than in any of those earlier sections, and that little mostly negative; for the biological field has been much less explored and examined by science, both theoretical and applied, than the material, and there is much less that the student of philosophy can take as premises for his own inquiry.

Biochemistry and biophysics are newcomers. They can already show great achievements in the more specialized and technical fields, and still greater discoveries may be close at hand. But on fundamentals they are still at the beginning. The biochemist finds that the human body functions like a laboratory—but automatic and continuously operating: choosing its own materials, accepting, or refusing and discarding; engaging in processes of greater complexity and delicacy than the best-equipped laboratories of our universities or colleges. The biophysicist, similarly, is exploring an elaborate electric installation, also automatic, which is dealing with the same kind of currents as he has been handling, but at a size-level about a million times lower.

When, however, either of them attempts to approach the underlying bases of all this, when he would explore life itself, or mind itself, the elements in the universe of which those phenomena are manifestations—what is he dealing with then?

He does not know.

For Bergson to speak of *l'élan vital*, or Bernard Shaw of 'the Life Force', is no help to us. Those are mere tautologies. The expressions that they use do not answer our question; they merely repeat it back to us. The mysteries of life remain—action, growth, reproduction, and mind. You know Tennyson's little poem:

> Flower in the crannied wall,
> I pluck you out of the crannies,
> I hold you here, root and all, in my hand,
> Little flower—but *if* I could understand
> What you are, root and all, and all in all,
> I should know what God and man is.

2

So we are brought to the great problem which has vexed the philosopher and the scientist from of old — the mind-body relation.

Here the chief scientific authority was — and still is — Sir Charles Sherrington. The most eminent neurologist of his time, both in this country and internationally, he devoted a great part of his long life to an intensive exploration of this problem, eagerly seeking the link that, in humans or other animals, enables bodies to affect minds and minds to affect bodies—as we see them doing every day and every hour. He accepted the identification of matter with energy, but when he came to the end of his long effort to find 'a scale of equivalence between energy and mental experience' he had to admit that his search had arrived at none. 'The two, for all I can do, remain refractorily apart. They seem to me disparate; not mutually convertible, untranslatable the one into the other.' He said again: 'Strictly we have to regard the relation of mind to brain as still not merely unsolved, but still devoid of a basis for its very beginning'. I had the great privilege of his friendship in the later years of his life—he was one of the most lovable of men. When I went to visit him in his retirement at Eastbourne, he was very willing to talk about these things. I well remembered his summing up his conclusion—we were walking one day, in 1946, along the seafront, when he stopped, and said with emphasis, 'It is perhaps no more improbable that our being should consist of two fundamental elements than that it should rest on one only.' In the concluding words of the Introduction that he wrote in the following year to the new edition of his principal work, *The Integrative Action of the Nervous System,* he said the same thing in the same words. And in his last utterance of all, when, in his ninety-second year, he recorded the opening talk in a broadcast symposium on 'The Physical Basis of Mind', his final words were: 'Aristotle, two thousand years ago, was asking how is the mind attached to the body. We are asking that question still.'

Lord Adrian, now the foremost figure among British neurologists, can say no more than this: 'I think we are a little nearer the threshold of mind than we were'.

3

In our present state of knowledge, then, it seems that—as regards both life and mind—we are at an impasse. The cytologists—now greatly aided by the electron-microscope—are busily at work on the living organic cell—its walls, its protoplasm, its highly complicated nucleus. It has long been known that electric charges and currents play an important part in the working of the brain and the nervous system; they have now discovered that certain amino acids seem to be an essential element of the cell. At any moment some team of chemists and physicists, in some part of the world, may hit upon a clue. But meanwhile we have none.

In this situation, some thinkers are content with the materialist philosophy which had set the tone for natural science during most of the nineteenth century. It was regarded almost as a matter of course, among the young people of my own generation, that, if one took a realist view of things, one must also be a materialist. There has been a great change since then, at all events in the countries of the western world where thought is free. To quote Sherrington again, I remember an incidental remark of his on this point—of which I made a note at the time: he said that 'the leading physicists of his day had been materialists, but he thought that was not so now. They had been so intent on atoms and the like that they were not interested in the problems of mind.'

In Russia this change has not been allowed to manifest itself. An incident in Moscow not long ago, which I have quoted elsewhere but which I would ask leave to recall here, is significant. A party of British Members of Parliament were there on a goodwill mission, and on a social occasion made the acquaintance of several of the principal Russian leaders. Mr Gromyko was discussing Marxism with one of the Members, Mr Christopher Mayhew, and asked him: 'But what is your own philosophy? Are you a materialist? Is that glass I am holding real or not?' Mayhew does not record his answer, but it might well have been—'Certainly I am a realist; but certainly also I am not a materialist': and he might have added another question in return. 'Your glass is a reality; but is not this conversation between us also a reality? It is not material physically—apart from the fact that it is conveyed to and fro by sound-waves in the atmosphere. But the ideas that we are exchanging—Marxism and anti-Marxism, communism, and politics in general; religious beliefs, and disbeliefs; my philosophy, which you ask me about and your philosophy, which you think better than mine—are not all these real happenings in a real world; just as real facts of the universe as your glass, or its vodka, and your own body which has consumed it?'

Of late, however, in the opinion of some, the declining influence of materialism has been revived, and by science itself, by the invention of that wonderful machine, the electronic computer. Now, it is said, mechanism is doing things which hitherto only human minds could do: this may be a beginning; in a hundred years all the rest may follow. 'The electronic brain' is a phrase which has come into common use. The question is asked 'Can machines think?' and some who ought to know better answer in the affirmative. We are told that the computer also has 'a memory', not very different from the memory in a human or animal brain. But surely an electronic computer no more thinks than the machinery in an automatic telephone exchange thinks when it is connecting one subscriber's number with another: or a cash register in a shop is doing a sum in arithmetic when it is adding the amount of the last purchaser's money to the cash already in the till: or an alarm clock, set for 7.30, 'remembers' when 7.30 arrives and it duly makes its noise.

The science of these machines, cybernetics, has probably a future of great usefulness before it. But it has no more connection with the problems of life and mind than has any other form of engineering. The engineer thinks. His machine does not.

4

When the philosopher is asked, what is life? and what is mind? he can give no answer. Also, in our discussion of the ether you put to me the question, what is energy? I did not take up that question then because I thought it would come up preferably in the context we have reached now: and now I can offer nothing better than the same agnosticism. The reason is that all three—life, mind, energy—are for us, in the present state of human knowledge, ultimates. We have therefore no words in which to relate them to anything that may lie beyond. We stand on what is, for the time being, the frontier of knowledge, and when we reach the edge of knowledge we have come to the limits of speech.

Here—and perhaps only here—the group of Oxford philosophers, who in recent years have urged that linguistics should be given a leading place in philosophy, may command our support. Otherwise we cannot accept a principle which implies that the best philosopher would not be the one who could best cope with the great problems of man and his place in the universe, but the one who could compile the best dictionary of philosophical terms. When, however, we try to pass beyond the phenomena which we can perceive or about which we can draw reasonable inferences, then indeed we are in the legitimate realm of linguistics. We have to consider whether our failure is due to the lack of words in which to express our ideas, or to the lack of ideas themselves.

Rousseau said that 'definitions might be good things, if only we did not employ words in making them'. Life, mind, energy—we know that they exist because we see all around us the phenomena—processes, events, objects—which they produce. We can see what they do, but we cannot say what they are. The fact that we cannot describe them in the language belonging to our man-centred world need not lead us to conclude that these are not real elements in the given universe.

5

To those three I think we are obliged to add a fourth. The word 'supernatural' is used in two different senses. In popular parlance it means something that is miraculous—that is to say, contrary to the laws of nature. A person who is dead is restored to life; a wooden staff is turned into a living serpent; a liquid, which at one moment would, if analyzed, show one chemical composition, at the next would show another. Often such events, accepted at face value by an uneducated and credulous populace, may afterwards be found capable of a rational explanation: but if not, they are recorded for posterity as events that are supernatural. Similarly with stories of ghosts, or other apparitions, or hallucinations.

It is difficult to avoid the use of the same word in philosophy when we wish to consider possible aspects of the given universe which are above—or outside—our own perceptions of nature. We cannot refuse to recognize that that world of ours is incomplete: it does not explain itself: it cannot have created itself. We are bound to infer that there must have existed, and must now exist, Something else—not included in the universe which we know and which we describe poetically as 'nature'—a Supernatural. Men have known it ever since, in the long epochs of the Stone-Ages, they began to watch the stars—and to wonder.

What it is has ever been a matter of speculation and argument. In the western world many—perhaps most—philosophers of the present day accept it as an abstract entity and name it Deity.

The human imagination would clothe this abstraction with attributes, but, having no language in which to describe it, has attached to it, from the earliest times of which we have record, attributes of honour or majesty such as those in which men had been accustomed to address their own kings or ancestors, but magnified and exalted. Then theologians and philosophers devise metaphysical terms into which to translate those ideas. They speak of Creation, of Omnipotence, Omnipresence and Omniscience, of Infinity beyond Space and Eternity beyond Time: philosophers may designate it The Absolute. But if we reflect upon those words we are bound to admit that none of them evokes in the mind any ideas of actual processes at work or events happening in a real cosmos. We are left where we were, with an abstraction and its name—Deity.

6

In this situation, men have had recourse to one or other of two expedients, or a mixture of the two—to symbolism or to myth.

Religious symbolism takes spiritual values and translates them into terms of human qualities. An abstract 'Deity' becomes a personal 'God': we no longer speak of 'it' but of 'Him'. The theistic religions evolve, bringing in the relationship of Father and children, King and subjects. The founders of the great faiths formulate the moral law—defining for the individual what is virtue, and for the nation what is righteousness. For centuries these creeds have commanded the reverence, though not always the obedience, of the masses of mankind. They have inspired the prophets and the saints. Numbers of individual men and women, in all parts of the world and in all ages, including the present, have held or now hold a binding conviction of the existence of a personal God with whom they, as persons, can hold communion.

Now between symbolism of this sort and myth there is a difference which must be clearly grasped and recognized as fundamental.

The essence of symbolism is that it is attached to something which those who employ it believe to be real—a factual element of the given cosmos. A myth, on the other hand, is not attached to anything; it is a work only of the human imagination. It may be supported by intuitions, or by aesthetic or other emotions, or by subconscious impulses derived from inheritance, early training and habit, but it is not supported by reason. The symbol is intended to represent or embody a truth: myth is, by definition, the antithesis of truth.[1]

Convinced as we are—if any body of constructive thought is to be evolved for the men of the future—of the necessity of the co-operation, not only of philosophy with science but of both with religion, we cannot be indifferent to this question of myth. For it is not to be denied that, in varying degrees, all the ancient theistic creeds still active in the world include, embalmed in their orthodox dogmas, relics of the old mythologies. Some have been included from the days of their foundation—ancient superstitions which were tolerated so as to make easier for the pagans the transition from old traditions piously held, and old ceremonies piously observed, to the doctrines and ceremonies of a

---

[1] A simple illustration may be taken from the Supporters of the Royal Coat-of-arms with which English people are very familiar—the lion and the unicorn. The lion is a symbol—of strength, courage, dignity: the unicorn is myth—no such creature as a white horse with a single straight horn (properly belonging to the narwhal) now exists or has ever existed. You can see a lion in almost any zoo, but no one will ever find a unicorn there.

new Faith of a higher order. And all through their hundreds or thousands of years of history, new myths have often been seeping in. When today intelligent and honest minds—especially among the young coming fresh to these issues — discard the myths, they often throw out religion as well.

Here we may find one cause of the relaxation of moral standards, which many discern among the younger generation of the present day, for religion has ever been a buttress of the moral law. But if that religion is based on myth, when the myth is detected and discarded—as sooner or later it must be, and in an age of widespread education it is likely to be sooner rather than later—then the buttress is destroyed and the whole structure may collapse. For, as Dean Inge wrote, 'The healthy human intellect will never believe that the same proposition may be true for faith and untrue in fact'.

The theologies are not so rigid as they are often supposed to be. The orthodoxy of one century is never quite the same as the orthodoxy of the previous century, and may be quite different, in important particulars, from the orthodoxy of five hundred years before. Religion does evolve. Those sincere and zealous adherents of the old dogmas, who today cling tenaciously to the mythical elements that still survive in them, are rendering ill service to the cause for which they are so deeply concerned. For changes, if they are to come, must come from within: they can never be imposed by legal or any other kind of compulsion from outside. If myth is still to be sanctified, religion itself may be lost, and the responsibility will be theirs.

But if myth is eliminated, then—and only then—the men of religion may be able to offer to mankind Faiths—divergent from one another no doubt in many points, and not losing their own savour and their own cherished traditions of heroism and sacrifice—but Faiths capable of holding their own in a scientific-minded age. Then, and only then, will they be able to preach creeds that can satisfy the religious spirit without alienating the intellectual conscience. Then and only then, will philosophy, science and religion find their meeting-place.

PROFESSOR DINGLE:

1

I agree with you entirely that there is a sense (though probably a temporal rather than an eternal one) in which the problems which you describe as those of life and mind are more important than those of matter, and that it is here that there is at present the greatest need to avoid strife between science and philosophy. These problems are also much more difficult: that is why modern science began with the physical sciences and why these are so much more advanced than biology and psychology. I need hardly say that it would be a presumption for me, who am no biologist or psychologist, to express an opinion on any particular problem in these sciences, and I shall not attempt to do so. But I do not think it is *ultra vires* for a physicist, from the experience which the highly developed state of his own science affords, to make suggestions concerning the means of approach that are most likely to bring success in other fields, and that is all I propose to do here.

Life to the biologist and mind to the psychologist are what matter is to the physicist. Before the seventeenth century, matter was studied as matter. The alchemists, for instance, sought to change lead as a whole into gold as a whole: they did not analyze lead—still less a particular sample of lead—into the sources of a grey colour, a hard feeling, a heavy weight, and so on. They met with no success. Progress in the physical sciences began only when, in effect, this analysis was made. The various *properties* of a particular piece of matter—its colour, motion, temperature, and so on—were studied separately, just as though the others did not exist. Thus we had sciences of mechanics, optics, acoustics, heat . . . each starting from its own basis without any reference at all to the others.

In the course of time some of the sciences merged together. The science of heat, for instance, combined with the science of motion into *thermodynamics*. Acoustics also became absorbed into the same more comprehensive science. The science of magnetism combined with that of electricity to form *electromagnetism*, into which optics eventually flowed. None of this would have been possible if the initial separation had not been made.

It would have been hopeless in the seventeenth century to attempt a science of thermodynamics or electromagnetism. Each science had to come to maturity by itself before marriage was possible. In the physical science of the present day 'matter' has no place at all, except as a word convenient to use when non-technical statements are being made. There are no laws of matter;

there are laws of thermodynamics, of electromagnetism, of gravitation.

It therefore seems to me that if we concentrate our attention on 'life' and 'mind', regarded as indivisible objects presented to us for examination, we are inviting failure. The objects of our study should be the separate phenomena which together are regarded as manifestations of these supposed entities. It is not for me to enumerate them: I want only to present a picture of the scientist or philosopher facing his problem, and that picture as I see it represents a thinker confronted with a whole mass of originally unrelated experiences which he can classify into experiences of various kinds. No particular classification is forced on him; he makes that which affords him the best prospect of finding regularities. He has selected a few classes—those already mentioned, which we regard as belonging to the physical sciences—in each of which he has found laws expressing the characteristics of the phenomena, and he has then been able to combine some of the classes into more comprehensive ones. Other classes have been selected, with less assurance, as the fields of biological and psychological sciences. The ultimate aim is to unite all experience under a single set of laws.

Consider now a human being as an object of study. This presents a mixture of almost every kind of experience—sights, sounds, movements, emotions, and so on. Some of these are dealt with satisfactorily in the physical sciences: the fact, for instance, that the man does not long remain without visible support fits in perfectly with the law of gravitation that applies equally to him and to a stone. The beating of his heart is partly included in the physical sciences and partly not; it belongs tentatively to some department of biology. His benignity or irascibility is partly physical, partly biological, and partly psychological. The physical parts of these experiences we understand (i.e. can express in terms of regular laws) pretty well; the other parts less well or not at all.

But these divisions are of our own making, and are subject to change as occasion may demand. Even in the physical sciences we see fluctuations of classification. It seems a single problem to ask how a piece of metal moves in the neighbourhood of the Earth, yet we find that iron moves in one way and copper in another. We do not invent complicated laws to cover both: we analyze the movements into two components, which we call *gravitational* and *magnetic*. The iron exhibits both, and what we see is a resultant of the two: the copper exhibits only the first to any appreciable extent. Again, the appearances on a cinematograph screen may be indistinguishable, as experiences, from those of people walking, yet we do not put them into the class of movements but into that of the appearances studied in optics, which are subject to quite different laws from those of motion.

Coming back, then, to our human being, we see that his movements—which, taken as the resultant actually presented to us, are quite outside interpretation in terms of any sort of regularity we can yet conceive—may be analyzed into components which do show regularities. The laws followed by those component movements may be as different from one another as the laws of mechanics differ from those of optics, or even much more so—they may involve concepts such as hopes and fears, 'desires and adorations, winged persuasions and veiled destinies', and so on, instead of mass, space, time and magnetic fields, and much of what we now think of as physical may turn out to be more properly described as biological or psychological—and *vice versa*.

The point I am trying to make is that the division into body and mind, or body, mind and soul, as separate entities presented for examination, is quite out of keeping with the character of science. There are no such entities. What we have to study are the phenomena we experience, and we can classify them as we like in order to get the most satisfactory system of regularities. The actual problems, of course, remain. We have still to understand why certain sounds make the blood rush to a man's face, but we shall not do it by trying to find how the changes produced in his 'body' by the impact of the sounds cause a disturbance in his 'mind' which reacts to cause a flow of blood in his 'body'. We may be able to express it in those terms, if we wish, after we have solved our problem, but to approach it in that way would be like trying to find out why gold, being heavy, is also yellow. We should pursue the study of emotions as one subject, and that of blood movements as another—as, in fact, I believe we do—and then, in course of time, we may expect their relation to one another to transpire, just as the science of thermodynamics reveals the relation between the specific heats of a gas and the velocity of sound in it. No direct attack on that problem would have been of any avail.

For this reason I would, with all respect—and I feel it almost an impertinence to say that my respect for Sherrington both as a man and as a scientist is immeasurable—suggest that his attitude to this problem was mistaken. The problem is, of course, a real one, but its character is obscured when we describe it as the 'mind-body relation'. If we persist in asking Aristotle's question I see no reason why we should ever stop. It is certainly 'no more improbable that our being should consist of two fundamental elements than that it should rest on one only', but what reason is there to suppose that it 'rests on' any 'fundamental elements' at all? A piece of iron does not rest on gravitation and magnetism, no matter whether these are conceived as independent, as they are now, or as different aspects of something more basic, as seekers after a 'unified field theory' hope they may be. What we have to understand are the movements of the iron, not the entities which we ourselves have postulated to help us in that task. And similarly, we want to understand our behaviour, not the relation between 'body' and 'mind' which we have ineffectively invented to serve that purpose.

2

In view of what I have just written, you will not need me to say that I share your rejection of materialism, and I see nothing at all paradoxical in the statement that a man may be a realist without being a materialist. I said in the earlier part of our discussion that I would not accept the label 'idealist' because of the misunderstanding and misrepresentation that would certainly follow, and the same applies, of course, to the label 'realist'. But inasmuch as everyone who tries to philosophize must have something to which he attaches fundamental importance and may therefore be said to regard as 'real', there is a sense in which we are all realists. What I regard as real is experience. You go further and regard the *cause* of experience as real—'the universe as it is'.

There, for reasons which I have given, I cannot follow you, and so, while to a certain extent I share your reaction to Mr Gromyko's remark about the glass he was holding, I would in this instance go further than you and say that the conversation was more real than the glass. The conversation (i.e. the hearing of the sounds, not the interpretation given to them) was a direct fact of experience: the glass was a hypothetical entity postulated to account for certain visual and tactual experiences. It is possible that with further experiences such a postulate might be found inappropriate and rejected (even in ordinary life; as I have said, it has long been rejected in science, where the visual and tactual experiences have been dealt with independently). For instance, if Mr Gromyko's experience had changed abruptly to something quite different, he would then have found it convenient to reject the hypothesis of a glass, and called his experience part of a 'dream' instead. But he would have no justification at all for doing this except that otherwise his other experiences would not make sense. The experiences formerly attributed to a glass would still remain unchanged, but they would be classified differently. Although exactly similar to experience of a 'real' glass, they would be given to psychology to deal with instead of physics.

This cannot be satisfactorily expressed in terms of 'reality' and 'illusion' if the former is regarded as pertaining to 'the universe as it is' and the latter to our 'imagination'. No one can possibly tell that what he is experiencing now is real in that sense or not: he may at any moment 'wake up'—or he may not. But what he does know is his experience, and he is at liberty to classify that as he chooses in order to rationalize it, and to change the classification for the same end with equal freedom.

The problem of the electronic brain is therefore simply this: can I make sense of things by placing human brains and machines in the same class and interpreting them in terms of the same concepts? It is a technical question on which I am not qualified to express an opinion, but while there are certainly more phenomena that are potentially common to both than would have been thought possible a generation ago, I should be exceedingly astonished if there did not still remain some aspects of human brainwork that have no analogue in any machine that can be produced.

We must, however, beware of false thinking in this matter. What anyone automatically tends to do when he compares a 'human mind' with a machine is to take himself as typical of the 'human mind'. But that is not legitimate. Each person bears a special relation to his own mind which he does not bear to the minds of others (I am, of course, speaking now in ordinary 'realist' terms to avoid the insufferable circumlocution that expression in terms of basic experience would entail, in a language not designed to conform to it: any inconsistency that may appear with what I have said about the irrelevance of 'mind', etc. is superficial and will not, I think, mislead any non-quibbler), and it is only the relation which he bears to the minds of others that is properly comparable with the relation which he bears to the machine. Consequently it is misleading to say that the engineer 'thinks' and the machine doesn't. We *say* that the engineer thinks because that is the hypothesis we consider necessary to explain our observations of his actions, but we have no direct knowledge of anything in the engineer comparable with the thinking we know in ourselves. Hence while each of us who is self-conscious may justly say that the machine is of quite a different character from himself, he is not on the same grounds entitled to say that the machine is of a different character from another human being. That may be true, but, if so, it is only to be established by showing the impossibility of making sense of things without keeping the phenomena associated with brains and machines in different classes employing different concepts for their rationalization.

I have been told of a machine for playing chess that, rather than lose, cheated. I am not sure whether that makes it more human or less.

3

I agree with you that *life, mind* and *energy* are words of the same character; where I differ is in the status assigned to words of that character. You regard them as denoting what are, at present, 'ultimates', representing other things that 'lie beyond'; these tentative ultimates are the representatives, in the 'universe as it is for us', of those other things that exist in 'the universe as it is in itself'. I see no reason for this duplication. Life, mind and energy are three words that we have created; they are not inescapably presented to us for study. To find out what they mean, therefore, we must ask ourselves why we have invented them; that, and that alone, can give their true meaning. The answer to that question is that we have invented them to stand for concepts that we have formed in order to make sense of our experience, and their meaning lies in the function which those concepts perform in that endeavour.

A very young child or a primitive savage would have no conception of what we mean by these terms: nothing corresponding to them would be presented to him in his daily life. Yet he would have experiences, very vivid ones, and he might also have a sense of wonderment which, if in time he became a philosopher, would inevitably impel him to postulate entities serving the purpose of bringing his various experiences into some sort of rational relation with one another. Life, mind and energy are names given to such entities which have actually been postulated in the course of philosophic and scientific thought. We have made them, and it seems to me a form of idolatry to look for some 'reality' which they represent. The time may very well come, if it has not come already, when we shall discard them, just as we have discarded phlogiston and caloric and telegony and trepidation and the celestial spheres and a multitude of other similar, though perhaps less comprehensive, concepts.

I am therefore inclined to agree with you that linguistic philosophy has some relevance here. My objection to it, like yours, arises from its overweening self-esteem (I am speaking metaphorically of the philosophy, not literally of the philosophers). I do not question its importance, but I am astonished at its assumption of all-importance. It is a means that proclaims itself an end, a tool posing as a masterpiece of creation. When a physicist makes an experiment, nothing is more important than that his galvanometer, say, shall give true readings, that it shall be calibrated, its zero error determined, and so on. If that is not done, and done thoroughly, his results are worse than useless: it is impossible to exaggerate the importance of this part of the experiment. But he does not write long papers about it. He does it as a matter of course, and other physicists take it for granted that he has not omitted this prelude to his work. If he did nothing but this, no one would be in the least interested. It is what he does with the galvanometer afterwards that contributes to physics. The linguistic philosophers examine words and sentences—which, like galvanometers, have been made by men for certain purposes — as though such an examination were the achievement of those purposes. It is difficult to see why one should trouble to perfect a tool and then refrain from using it. I hope I have shown that there is scope enough for a critical examination of the ideas behind the statements of modern scientists, but this no philosopher comes forward to undertake. He carefully enumerates the various meanings which a statement can bear according to the rules of grammar, and then leaves the meanings unexamined.

4

'The supernatural' is on rather a different footing from other terms under discussion. As you say, it is used in different senses, and one hesitates to speak about it because of the probability that meanings not intended will be read into whatever one says. However, risks must be faced, not shirked, so I will be as unambiguous as I can and hope for the best.

By 'nature' you mean, as you say, 'the universe which we know', and by supernature you mean, not merely another part of the universe, of the same character as nature but which we do not know, but rather something of a different kind in which it is necessary to believe in order to explain how nature came to exist. That seems to me to lead to an endless regress: the same reasons that demand a supernature would demand a super-supernature to explain how supernature came to exist; and so on. But the problem does not arise for me, for I do not accept 'nature' in your sense, so it is impossible for me to be impelled towards supernature. My equivalent of your 'nature' is experience — that which is presented to me for study. It seems to me the inescapable starting-point of philosophy, for even if you postulate an external universe, all that we can discuss concerning it is the aspect which it presents to our experience; anything outside that must necessarily be inaccessible to us.

The use of the word 'supernatural' in connection with the so-called miraculous raises different issues, for here there is no question of going outside experience; a miracle that we do not experience is a sheer fantasy. By definition, on which we doubtless both agree, a miracle would be something that we experience that runs counter to the regularities detected in our normal experience. Let us take as a type what seems the most unlikely of your examples—a wooden staff is turned into a living serpent. Suppose we observed this actually to happen: how could we infer from it that there was a supernatural agency at work? Only by first accepting the view that there is an external something called nature which takes a course controlled by invariable laws, among which is one that makes it unnatural for wooden staffs to turn into living serpents. That is to say, before the supernatural can be affirmed you must accept not only the existence of the natural but also the finality of the laws now regarded as governing the natural; otherwise it could be maintained that the occurrence was perfectly natural but that the law under which it occurred was imperfectly understood.

I was on the point of saying that probably no one at this time of day, knowing how scientific history is strewn with the corpses of 'eternal natural laws', would maintain that our present ones are final, but then I remembered the 'creation-out-of-nothing' theorists, so I will leave that point. To me 'miracle' is a useless word. If I am forced to use it I am equally ready to say either that we have no reason to believe that there is any such thing or that most of our experience is miraculous. At the beginning, everything that happens comes like a bolt from the blue: we did not expect it, and we have no inkling of what will come next. Gradually we detect regularities, and then we are able to state 'natural laws' which tell us in advance what will happen if we do certain things. We cannot possibly know that these laws are infallible or that there are no alternative laws that cover past experience equally well but give different prognostications. The formation of shadows was explained perfectly satisfactorily by the particle theory of light, but the prediction of that theory concerning the change in the velocity of light when it passed from air into water was violated. In the absence of an alternative theory this could have been called a miracle, but in fact another theory was adopted that explained the new fact and also the formation of shadows.

Suppose, however, that no new theory is forthcoming. We can then either regard the existing theory as 'right' and the violation of its prediction as a 'miracle', or else say that the refractory occurrence is just another of the many experiences that we have not yet succeeded in bringing into rational relations with the rest of organized experience, and go on trying to work it in. I have no hesitation in choosing the latter alternative. We are at present at some point in the long process of rationalizing our experience at which we have much success behind us and much more unsuccess before us. The history of science is the growth of the former at the expense of the latter. It does not seem to me that I am saying anything very profound if I describe the appearance of a light when I depress a switch as a natural occurrence because I could forecast it, and the appearance of a meteorite in my garden as a miracle because I could not forecast it. All that is relevant is summed up in the statement that some experiences have been related together and some have not, and that there is every reason to believe that the former quantity will grow. I do not expect to see staffs turn into serpents, but also I did not expect to see a photograph of the far side of the Moon. If I should ever get the former experience I should investigate to see whether I was drunk or hypnotized or wait to see if I woke up or something of that kind, and, if the result were negative, leave the matter as so far unexplained.

All this has nothing whatever to do with religion as I understand it. That is an entirely different matter, to which I now turn.

5

I enter on this subject with great hesitation and with a full consciousness that anything that I can say about it is utterly inadequate, uncertain and, as likely as not, obscure. With negligible exceptions we all see and hear, and describe what we see and hear in compatible terms; hence we can talk about such experiences with little fear of being at cross-purposes, and accordingly the physical sciences have made great progress on which we can all agree. The experiences with which we are concerned in religious matters are quite other. To some they are far more intense than anything else; to others they seem scarcely to exist; and between these extremes probably all intermediate stages are represented. Even among those to whom what is called 'religious experience' is a vivid reality, very imperfect agreement is found in their accounts of it, and I doubt whether it is possible to describe some forms of it in such a way as to distinguish them from experiences induced by drugs.

It will be at once evident why my approach to this subject must be quite different from yours. You have introduced religion by way of a postulated entity in 'the universe as it is'. What we experience must have a cause; that cause is unknown—we can only give it a name, the supernatural, Deity. That name symbolizes the reality beyond, which is the object of worship of the religious man.

That course is not open to me. I must start with experience, and anything that I postulate in order to make sense of experience—which would be my equivalent of your entities in the universe as it is—is my own introduction into philosophy and cannot possibly be an object of worship. I could no more think of worshipping any Deity arrived at in that way than of worshipping energy. Such a Deity would bear exactly the same relation to the experiences which we call religious as energy bears to the experiences which we call sensory.

I must start with experience. Speaking in the broadest possible way, I would say that three attitudes are possible, and are practised, towards experience taken as a whole. They are not mutually exclusive; they are all possible—though rarely, if ever, found—in the same person. First, one may be satisfied with the bare experience, content to enjoy it for no reason beyond itself. This attitude is represented in many types, distinguished by the kind of experience which they seek: the athlete, the sensualist, the sadist, the masochist, and, with a qualification to be mentioned shortly, the religious mystic—these among others come under this heading. Secondly, one may feel an urge to express the experience in some form—in music, painting, poetry. The general name for such a person is 'artist'. The symbols of which you have written are of the character of artistic creations, though they may be good or bad, profound or trivial, art. (I agree, by the way, that a myth and a symbol are distinguishable, but I do not think the distinction is quite so absolute as you have described it. A unicorn is a myth in a sense in which a lion is not, but in the coat of arms I think they are equivalent as symbols, though I am not very clear what they symbolize. But I do not see how the symbolism of the lion could have any meaning if the unicorn does not also symbolize something related to it.)

The third attitude to experience is that of which we have been talking most of the time, which leads to the attempt to form it into a rationally connected system. This is the attitude of the scientist and the philosopher. Unlike the others, he does not select: the whole of experience is his field. Those who simply enjoy or express experience choose what experiences they like and leave the rest: the scientist or philosopher finds himself committed to the whole. For practical reasons he may specialize—the physicist need not be a psychologist—but his subject necessarily spills over into other fields: even in physics, for example, we must take account of 'personal equation', which is a psychological phenomenon. This universality is more conspicuous in philosophy than in science — at least it was before philosophers betrayed their calling by exchanging experience for elementary creations of the artist — words; but the distinction between science and philosophy is not deep-lying and is usually (though, as I may suggest later, not necessarily finally) best ignored.

The essence of religion, as I understand it, is in what I will here temporarily call 'religious experience', and the religious man is he who enjoys it. He may also express it and try to rationalize it, but that is secondary. If he knows what it is he has the essence of the matter, even if he cannot express it in any form or give assent to any rational creed. On the other hand, a man may paint 'religious' pictures which are masterpieces, and believe every word of the Apostles' Creed, and yet have no religion in him. But, unlike any others who adopt the first of my three attitudes to experience, the religious man, in my sense of the term, does not enjoy it for its own sake alone, for by religious experiences I do not mean those mystical experiences that come to few, but appear, when they do come, to overshadow all else. Such experiences, when they accompany what I call religion, may be a great blessing, and otherwise a great disaster. I mean by religion the much commoner state of consciousness, often called prayer, in which one takes a course of action to which he feels impelled but for which he feels incompetent, relying on an accession of strength beyond that normal to him. Religion in this sense is therefore essentially related to action, and therefore to morality, which I shall consider later: by definition it cannot exist for its own sake.

I am purposely avoiding such phrases as 'relying on God' because they take one outside religion into a creed, however rudimentary, and that at once introduces all the problems over which men have argued and fought for centuries. I have no doubt whatever of the 'reality' of religion, and in saying this I am not appealing to my own experience, which is of no importance to anyone but myself, but speaking quite objectively in the light of overwhelming evidence. I must believe either that large numbers of men and women, who appear to me to be among the most admirable of their kind, have had this experience, or else that they are liars. I cannot help choosing the first alternative. The fact that others, who may be equally admirable, say that they have never had such experience does not prohibit this choice. I cannot deny the 'reality' of sight because there are blind people, nor could I do so if blindness was far commoner than it is.

Of course, it is also a fact that most, though not all, religious people do associate their religion with a creed — in its most elementary form simply with a belief in the existence of God. Whether for such people on the whole this is a good or a bad thing I do not know, though, for reasons which I shall give later, in a final summing-up, I think such an association (I am not speaking of the things associated) is inimical to a satisfactory reconciliation of religion with science. The moment you attempt to rationalize your experience by postulating an entity responsible for it, you must give some description of that entity, otherwise you have done nothing more than said, in less direct terms, that you have had the experience. And immediately you try to describe God you inevitably bring in elements which are not justified by the experience that leads you to assert his existence. The question is therefore: is the harm done by the introduction of these extraneous elements worth putting up with for the sake of the greater sense of 'reality' that comes from substituting a transcendent Cause of the experience for the experience itself? I cannot answer this question. The 'realist' attitude, in your sense of the word, is so deeply rooted in our thinking that to most religious people the religious experience seems identical with belief in God, and they could not forgo one without losing the other. On the other hand, when I think of the tortures and persecutions and wars and intolerances that have characterized the history of 'religion' through the centuries, up to and including the present, and remember that these are due entirely to differences in the inessential concepts that have been introduced with the word 'God', I feel that the problem is beyond me.

The trouble is that the word God is used in two senses, which I may call, merely for the immediate purpose of distinguishing them, the religious sense and the scientific sense, and these are so inextricably confused that most discussions of the problem are foredoomed from the start. When we say that God exists in the religious sense we should be stating nothing more than is justified by the fact of religious experience. But we often reach a conclusion that is expressed by the same words from a consideration of ordinary sensory experience — i.e. from the facts of science—and then the God who is our refuge and strength is identified with a hypothetical agency responsible for the origin of matter and for various other things which have really nothing at all to do with religion. In that way we are led into contradictions which we disguise to ourselves as 'problems' and make the subject of endless debate.

The 'problem of evil' is an outstanding example. God has made the universe as it is, and He is all-loving. But the universe is full of cruelty: how can that be?

> This good God—what He could do, if He would,
> Would, if He could—then must have done long since.

This is not a problem to be puzzled over; it is a plain contradiction, which means that the premisses cannot both be right. No amount of special pleading can alter that simple fact, yet we go on hoping that somehow the impossible will happen.

All this waste of time is avoided when we go behind the hypotheses, the creeds, into the experiences which are the evidence for them. The evidence for a God of love lies in religious experience; the evidence for a God of power lies in physical experience. There is no contradiction between religious experience and physical experience because in fact we have them both, and that settles the matter. Hence the contradiction must have been introduced by us in making our inferences. Either we have falsely inferred one or the other or both, or else we have wrongly identified them with one another. The position is just as though nineteenth-century physicists had inferred a wave-carrying medium from the phenomena of sound and also from the phenomena of light, and had given their inferences a single name. The discovery from acoustical phenomena that this medium could be removed by an air-pump would then have been contradicted by the discovery from optical phenomena that it could not be so removed. Thus would have arisen 'the problem of the vacuum'. It would have been in all respects similar to the problem of evil.

It seems to me that the primary attitude which we must adopt towards all experience is to accept it. I do not mean, of course, that we should not use such powers of control over things as we possess in order to produce the most desirable experiences and eliminate the others, but in our thinking we must accept the fact that experience is what it is and not try to explain it away. I do not, like Margaret Fuller, accept the universe—that I merely postulate for certain limited purposes—but I know that the only wise thing to do is to accept experience. Having done that I can form whatever hypotheses or creeds I find useful in making sense of it, but I must never let them tempt me to deny any experience that does not conform to them: the hypotheses and creeds must go, not the experience.

6

In all this one thing has been left out of account—the ethical problem. Experience comes to me from the mere fact of my being alive, and I am able to think about it and attempt to rationalize it from the fact that I possess reasoning power. That is all that is necessary to describe the process of philosophizing. But also, being alive entails the power of doing things, of interfering with the succession of experiences that would have come naturally and making them something else; and the more I succeed in making sense of experience, the greater becomes my power of ordering its course. I have therefore at every moment a choice of a large number of alternative ways of acting. How shall I exercise that choice?

It seems to me that Kant, in speaking of the guiding principles of thinking and of acting as 'pure reason' and 'practical reason', made them appear to resemble one another much more than in fact they do. For the outstanding fact is that, whereas I am naturally endowed with an intuitive knowledge of principles of pure reason, which I share with all other sane persons, I am not so endowed with a knowledge of the principles of right action; I have to choose them for myself, and when I have made my choice it does not agree with that of others similarly placed. All agree—they cannot help it—that, the definitions of the terms having been agreed upon, twice two are four, things that are equal to the same thing are equal to one another, and so on. All do not agree that war may sometimes be justifiable, that lying is sometimes necessary, that race segregation is a permissible policy. It is true that in some cases there is a conviction on these problems of action that seems to its possessor as indubitable as the axioms of reasoning, but that is not the general rule, and such persons are usually better described as bigots than as prophets. In every situation we must act (doing nothing is an action in this sense, since it entails effects that would not follow without it), but either we have no inner monitor to tell us how to act or, if we have, it tells different people contradictory things.

Of course we try to show that our particular ethical principles

are 'rational'—how many attempts have been made to do this!
—but if we are honest we must admit that all such attempts are
doomed from the start: if they could be successful, all reasonable
people would have the same code of morality, just as they have
the same multiplication table. To take a single example, I do not
accept the Nietzschean doctrine that everyone should act in such
a way as to hasten the coming of 'superman'. If one is of the
master type he should enslave the herd; if he belongs to the herd
he should either accept that status and act accordingly or, if he
is 'unfit', be eliminated. But if I am told that I do not accept this
doctrine because I am not of the master type and do not like the
subservient position allotted to me, I have no satisfactory answer.
I cannot deny that it is possible to read the course of evolution
in such a way as to show a continuous increase of power of some
creatures over others, and if it is held that it is the duty of every
individual to 'accept the universe', to align himself with the
tendency of the whole (a view held also by non-Nietzscheans and
often called 'submitting oneself to the purpose of God as revealed
in nature and history'), I can offer no convincing refutation. In
the last resort I can only say: 'I reject it. I cannot justify my
rejection, but I find myself naturally responding to a different
ideal, and I yield to that response.'

There are not many Nietzscheans, I believe, but the same
principle applies to every moral question. Everyone's convictions
on such matters rest ultimately on this dogmatic basis; and on
many of the most important issues, of both individual and corporate action, we have the spectacle of the most estimable and
conscientious people holding contrary views, each with the unshakable conviction that he is right. (There are also those who,
realizing the complexity of the problems, cannot form any convictions at all; but our space is limited, so I pass these over.) This
is an undeniable fact: what are we to conclude from it?

In the first place, it seems to me to rule out the theory that
there are certain absolute moral rules applicable to everyone—in
a common phrase, that there is a unique 'will of God' for every,
or indeed any, situation. If that were so, then, in every controversy, one and only one of the disputants could be 'right' and
all the others must be 'wrong'. We should have to say of Luther
and Loyola, to take but a single instance, that one at least should
have acted in the opposite way to that in which he did, which
means either that he was a conscious impostor or else that it is

one's duty to act contrary to his deepest convictions. I cannot accept either of these alternatives. There are so many instances of the kind that the former strains my credulity beyond breaking point and the latter would make havoc of all morality; our convictions would mean nothing since we should never know whether to follow or to violate them.

Considerations of this kind lead me to the conclusion that the moral conflicts evident all through human history are ultimately to be seen not as a battle of right against wrong, in which the important thing is that one side (the right) shall win, but rather as an interaction of morally equivalent opposing elements leading to something not identical with either but which would not have been achieved without both. 'Satan' is a conspirator with 'God', as in the Book of Job. The best simile I can think of—greatly simplified and inadequate, but still helpful I think—is that of a football match. It is right for A to do one thing and for B to oppose him; it would be wrong for A to do what is right for B. The important thing is not that either A or B should win but that there should be a good game, and there will not be a good game unless each tries his utmost to frustrate the other. God—in the sense in which I have used the word, i.e. not including anything unauthenticated by religious experience—is in part the referee, but this is a very inadequate image, for He is available to help both equally, and, in fact, the more they rely on Him the more effective their efforts become.

It is very difficult for an earnest person to believe that what seems to him to be wholly and undubitably right is not 'God's will', and what seems wholly and indubitably wrong is not 'contrary to God's will', but the facts seem to me to make that inescapable. The solution to one's moral problem is not to find the unique 'right' thing to do, but having found what is right *for him* to do, to do it with all his might, in the conviction that the outcome, which he cannot possibly conceive or control, will be for the best—whatever that may mean. That conviction, I think, can only be a matter of faith. I do not know how it is to be obtained; I only know that I have it. For reasons of which I am unaware and therefore cannot defend, I believe that 'all is for the best'; the equivalent in your terms would be, as someone has expressed it, that 'the universe is friendly'. In so far as I do add my efforts to the support of what I believe to be right and the destruction of what I believe to be wrong, I feel a sense of satisfaction that does not come from any visible sign that my efforts are successful or will ever be successful, but from an inner conviction that I am taking my proper place in the general pattern.

It is tempting to try to find some external justification for this faith, if not in reason, at least in the course of history as it has so far proceeded, but I doubt whether this has much significance. I have sometimes wondered whether we are not so constituted that we automatically accept as 'best' what comes to pass, so that we ought to say that the best is what happens rather than what happens is for the best. Obviously this is not true in detail, but it may be so in the long run. We take it for granted that we are 'higher' than the apes. I do not know what the apes feel about this, or why, objectively, their view on such a point, whatever it is, should be given less weight than ours. But we have 'happened', and we are satisfied that that is 'good'. Even in comparatively trivial and short-term matters we could find support for such a view. In the nineteenth century many of the most conscientious people regarded the social equivalence of men and women, sex education, mixed bathing, sabbath 'desecration', free biblical criticism, as evils to be resisted. Now all these things are commonplace, and we regard them as blessings and ourselves as emancipated from crippling restrictions. No fresh *evidence* has come to light on the moral status of these things: the changes have simply happened through the interaction of a variety of influences not designed for that end, and, having come, they are welcomed as an advance. If we could foresee what the twenty-first century will regard as emancipation, might we not be surprised? And might it not be that in general we adapt ourselves to a course of development not of our devising? This is, of course, only a speculation.

The idea of individual moral standards which I have adumbrated, to which it seems to me that we are forced by facts quite independent of those dealt with in the earlier stages of our discussion, is entirely conformable to a philosophy based on experience as the primary 'reality', but hardly so, I think, to a philosophy that begins with an objective 'universe as it is in itself'. If there are any moral standards at all, one would expect them, on such a philosophy, to be enshrined in the universe and therefore to be the same for everybody. Either Luther or Loyola, if not both, should have refrained from his immoral acts, though how he would know this I cannot conceive.

LORD SAMUEL:

1

I am glad to find in what you have written some points at least in which we are agreed: I am not surprised that they should be few.

We have given a great deal of our space to the topics which dealt with problems in physics, and now we are proposing to give relatively little to this Section, which touches on problems in biology and psychology; but we are agreed that that does not imply that we value the importance of those subjects in the same proportion. On the contrary, weighed according to value in the long run to the well-being of mankind, life and mind—ideas, the humanities — take priority over matter, things, technology, economics.

We agree likewise, that if the word *realism* is to be used at all —and for my own part I regard it as a very useful word—it would be quite wrong to equate it, as is often done, with materialism. I rank myself as a convinced pro-realist, and as definitely an anti-materialist; and I should hold that the Communists, who at present control the policies of most of the countries east of the 'iron curtain', would render a great service to mankind if they were gradually to shift their social philosophy on to that basis.

I am happy to find also that we are both recalcitrant to the group of logical positivists who were able to assume leadership for a time among Oxford philosophers. Precision in the use of words is of course essential to philosophy, but to give to linguistics, as they would do, the central position in philosophy could not be more than a transient fashion.

2

This is, I think, a suitable place for a note that I should like to insert, but I offer it with hesitation. For it is not put forward as a proposition, or a hypothesis, and it can hardly reach the dignity of a speculation; it can rank only as a possibility. It may, however, be worth bearing in mind when we find ourselves standing at the present frontier of knowledge with a misty view beyond of vast domains to which we give such names as Energy, Life, Mind and Deity, but with no vocabulary with which to define them and no images to picture them.

The progress of philosophy, including science, has many times been held up by assumptions which seemed so obvious that they did not need even to be examined: for example, that the Sun moves round the Earth, or that a material object 'possesses properties' and that these account for its 'behaviour'; or that the ultimate constitution of matter is granular, consisting of indivisible atoms, of the same order as grains of sand. Similarly, when we come nowadays to discuss ether possibilities we assume, as a matter of course, that we are considering whether there exists 'an ether', and that is our starting-point. If there is any ether, that it must be 'the ether', the only one, is taken for granted. The possibility that, if there is one, there may be more than one, has never so far as I know, been envisaged. The question of 'the ether', has already caused so much trouble — especially since Michelson-Morley—that anyone who ventures to suggest that there may be two ethers, or even three or more, is likely to make himself very unpopular with physicists. And it must be at once admitted that there is no reason to believe that either philosophy or science can put forward any evidence in favour of an assumption of more ethers than one. But neither can they put forward any grounds for the present tacit assumption that, if any, there is only one.

In the universe posited by our mathematicians, there is plenty of room for such a speculation. Within ranges of their measurements there is a long way to 10^{-27} in littleness or 10^{26} in bigness yet the one is part of the formula for Planck's quantum and the other brings us, through the Palomar telescope, into the realm of the Milky Way and the other galaxies. Within those bounds there may be all kinds of things not dreamt of which have existence in their own right, and outside these limits conceivably many more. Suppose we could reach 10^{-50} or 10^{50}, what might we not find? We may do well to keep possibilities such as those at the back of our minds when we are confronted with the concepts of Energy, Life, Mind and Deity, and 'can see what they do but cannot say what they are'.

3

Coming to our immediate topic, I fully concur again on one fundamental point—where you postulate that life and mind are not 'entities'. We cannot avoid in ordinary conversation the use of those words as convenient—as we cannot discard such familiar words as space and time, or the Sun 'rising' and 'setting': still, strictly speaking, we ought not to think of a living organism having life because it 'possesses' a fraction of an element, to which we give that name, which is part of the given universe: and similarly with a thinking organism, and mind. But, so far as I can see, the remainder of your contribution is governed by the divergence between us which you emphasized at the beginning of our Dialogue, and which you foresaw would trouble us all along—as it has divided philosophers all through history. It is a divergence that, I contend, can only be overcome if we realize that there are those two fields of study which we have argued about so often—first, a cosmos which to us is given, which exists in its own right, as it is and for what it is: and second, a man-centred universe, which is our own production, and which it is our business to bring gradually into conformity with the other, taking that other as the standard of what is true or false. You, on the other hand, are equally insistent that, for human beings, there is no subject-matter accessible for our philosophizing except our own experience; and our only standard of truth is whether we can relate one part of our deductions from experience with other parts without finding ourselves involved in contradictions.

Necessarily, holding fast to that conviction, you can only approach from that standpoint what the ordinary man calls the problems of life and mind, as you have approached the problems of physics. But I am equally bound to refuse to submit to that limitation. For me every individual animal or plant organism presents the problem—What is Life? and every animal with a brain and nerve-system presents the problem, What is Mind and how is it related to the material body with which it is associated?

I would offer an illustration which I have given elsewhere.[1] I may be walking in the autumn in some park where there are chestnut trees and the chestnuts are falling, and I pick one up. Nearby I see what at first glance seems to be another chestnut, but which proves to be a pebble, of much the same size, shape and colour. I pick it up also, and let the two lie side by side on the palm of my hand. I consider them carefully: there is a difference between them which is fundamental in our understanding of the cosmos. For the pebble is what it now is, and,

[1] *In Search of Reality*, p. 62.

kept for a hundred years it will so remain. But the chestnut is not only what it is now seen to be. It has the capacity to become something else that is quite different. Let the conditions be favourable and it will germinate; it will put out roots and a stem, branches, twigs and leaves. It will gather materials from the soil and the water-vapour in the air, from the light and heat of the sun. In time it will become a great spreading tree, and that tree may produce hundreds of chestnuts, each one with the same capacities. I know that the one I am holding is composed of molecules, atoms and particles, and that the pebble is also so composed; though the elements are different. But the chestnut has—and the pebble has not—that other potentiality of growth and reproduction.

Similarly as to mind. You take as your example a human being, but I suggest that that is liable to import again the subject of human experience, and to confuse the issue with our old source of discord. It would be better to take some other example of a thinking organism. At this time of year—spring passing into summer—I can watch every day, from the window of the room where I am writing, the birds of many varieties which frequent the old garden of this house—conspicuous among them the wood-pigeons, which for some years have been nesting in the thick ivy round my window. Let me take one of these birds as my example of a thinking organism. It is busy in its own section of the universe, choosing its mate, deciding on a safe place for nesting, picking up the right materials for the nest, incubating the eggs, feeding the fledglings, encouraging them in their first attempts to fly. It may be said that all this is merely the product of inherited instinct, or of the subconscious mind, and not of conscious reasoning. That may be so; but the choice at each stage at the right time, followed by the consequent action, is undoubtedly mental. The bird is doing things which the ivy on the wall, or the plants in the flower-bed, cannot do. There is a basic difference between them, as there was between the chestnut and the pebble. They are the differences between life and not-life, between mind and not-mind. They typify differences that run all through nature. They present problems that the inquiring spirit of man cannot ignore and will not seek to evade. Sherrington was right to spend years of his life in the effort to solve them, and, when he had failed, was right frankly to say so.

Some will hold that it is a mistake to think that there are two separate problems, one of life and one of mind, but that both are of the same order; or if not of the same, perhaps at least of a similar order—wherever there is life there is also mind, no matter how embryonic, even in the tree, even in the single organic cell. They hope that, if and when the researches into the structure of the cell have revealed the structure of the atom, then the discovery of the nature of mind will follow, perhaps soon. But for our present purpose this point is not important. Whether we are faced by two differentiations—(a) between matter and life and (b) between matter and mind; or whether it is a single differentiation between matter on the one hand and life-mind on the other—this is of little significance for our general purview. Of great interest in itself, for us it is merely a question of procedure, and to pursue it would be an unnecessary digression.

4

Starting from your different standpoint, you begin by citing the failure of the alchemists in their attempts to turn 'base metal' into gold, and you attribute such failures to the early scientists not having analyzed the 'properties' of the metal into separate problems—'colour, motion, temperature and so on': a piece of lead, for example, into 'the sources of a grey colour, a hard feeling, a heavy weight, and so on'. And you say that this was wrong because the scientists ought first to have divided their subject-matter into separate studies, and classified the so-called properties into the materials for several sciences. Each one of these has the function of finding 'laws', which would govern our own experience — 'sights, sounds, movements, emotions, and so on'. Then these laws—of thermodynamics, gravitation, magnetism, and others—can be correlated with one another, and that is all that we can expect to be able to do, and therefore all that we should try or wish to do. So you conclude (p. 188): 'The division into body and mind, or body, mind and soul, as separate entities presented for examination, is quite out of keeping with the character of science. There are no such entities. What we have to study are the phenomena we experience, and we can classify them as we like in order to get the most satisfactory system of regularities.' You end by saying, 'The actual problems, of course, remain'. But you add that in the biological field, they are not such as can be denoted by the words, *body* and *mind*. Therefore you are unwilling to discuss Sherrington's inability, in our present state of knowledge, to find some kind of monism that would comprise both mind and body, and his reluctant acceptance of a possible dualism.

Starting from my different standpoint, I come to quite a different conclusion, to which I still adhere. I miss in your account of the situation one word, which for me is a key-word for this discussion: it is the word 'process'. I agree that a material object does not 'possess properties', such as the early scientists would ascribe to it, but there is another alternative besides the one you favour. I look upon all the phenomena as manifestations of processes that are taking place, all those processes engaging, not only the object, but also its environment.

The piece of lead is 'grey', because, from some source of light, a train of electromagnetic light-pulses falls upon it; and its surface is of such a character that, at a normal temperature, the metal will reflect sections of the light rays which are of a certain wave-length: it is that which conveys through our optical nerve systems to our minds the sensation we conventionally describe in English by the expression 'seeing grey'. The rest of the light spectrum, I would say, is stopped by the opaque piece of lead, and is ended: or as you would say, its energy is absorbed by it. Whichever it may be, if the event is happening in a room with an artificial light and the light is turned off, the lead ceases to be grey; it has no colour at all. This establishes that 'colour' is a phenomenon that is caused—and I stress the word—by a combination between an object which can reflect and light which is conveyed to it by means of some process of transmission that is going on in its environment.

The lead gives rise to 'a hard feeling': this is because it is solid, and it is solid because the atoms and molecules that compose it are made to cohere by electrical or magnetic forces which bind them together. These binding forces may vary in strength and the cohesion is consequently firmer or more relaxed.

If a piece of the lead is put into a crucible and heated, the atoms that compose it are engaged in a process of greater thermal agitation. This changes it also from solid to liquid, and its colour as well as from the grey wave-length to the red.

The lead is 'heavy': it has weight; and that means it responds, like all material objects on or near the Earth's surface, to the pull of the Earth's gravity. This is a continuous process which we have not even begun to understand. It seems to be a force which radiates outwards from the surface of the planet, but which, when it meets a material object, undergoes some kind of reflection, like light or radar, but which sets going a movement in the opposite direction, which, if the object is free to move, carries the object along with it. Whatever may be discovered to be its nature and its mechanism, gravitation is a physical process, characterizing our planet, as the other stellar bodies. This can now be proved to be so by observation and experiment, because if a piece of the lead were part of the equipment of an artificial satellite which is sent into orbit round the Earth outside the effective range of its gravity, it would have to be attached to the wall of the missile, because we know that it would no longer have weight but would float around at random.

So here we have five processes, operating in the lead together with its environment: (1) the reflection of light; (2) the binding forces that give internal cohesion to molecules, atoms and particles; (3) the thermal agitation of atoms; (4) the Earth's gravitation; to these we must add (5) its magnetism.

You have not attempted to present a complete list of what would formerly have been called the 'properties possessed by' the piece of lead and will not demur to my adding one more. If the lead were combined with other metals and used to make a bell it would then 'possess resonance'. But this also would in fact be a process—when struck by the hammer the molecules or atoms constituting the bell would vibrate, and this agitation would set up pulses in the surrounding air; these would travel some distance until they became spent: if meanwhile they were to affect the ears of a human being, or some other animal, or a radio record, we should call them 'sound-waves'. Here we have an example of a sixth process.

After centuries of laborious research, we have come to know a good deal about some of these processes, but about others very little. If an encyclopaedia of 1960 containing accounts of thousands of the successes of science, were to be compared with what would have been possible in a compendium of knowledge in 1460, we should see that, whatever may have been the mistakes of procedure that hindered achievement, mankind has been on the right track in concentrating on the direct investigation of nature's processes. Those processes are what they are as parts of a universe, which is what it is. They—and perhaps many others—go on everywhere and eternally, regardless of human experience or activities. Our classifications and definitions, our measurements, probabilities and mathematics in general, our speculations, hypotheses, theories and 'laws', have no effect upon them at all. We may be confident that, without them, a nameless Sun would continue to exist and go on radiating; that the atoms in material things on the Earth would go on reflecting the Sun's rays and being affected in variable degrees by its heat; that things would continue to adhere to the Earth's surface through what we call its gravity; that molecules, atoms and particles would exist, and binding electrical forces would still give them internal and external cohesion. At all events, we may regard the probability of this as so great that a philosopher may feel justified in basing his cosmology on that belief.

You say 'The actual problems, of course, remain' (p. 188), but a realist philosopher would not agree to leave the matter there. The great names of science would forbid it. For if it had been the rule to take human experience as our subject-matter rather than those processes and their like, then Copernicus, Galileo and Kepler could never have got started; nor Newton; nor Marie and Pierre Curie, J. J. Thomson and Rutherford; nor yet Darwin, Pasteur, Mendel.

As to the problem of the supernatural, and the two senses of that word, I do not feel that I could usefully add anything more to what I have said in my opening contribution to this section.

PROFESSOR DINGLE:

I think the only comment I can usefully make before we come to our summing-up concerns your suggestion of a multiplicity of ethers. You are quite right in maintaining the legitimacy of this conception, but it is not correct to say that such a possibility has never been envisaged. For instance, the idea of Faraday's which I have mentioned elsewhere (p. 65), of rays issuing from, and moving with, every atom or elementary particle, is, in effect, the idea that every particle carries its own ether about with it, so that there are as many ethers in the universe as there are particles. However, in that case there would be only one *kind* of ether; the various specimens of it would differ merely in motion and the position of the centre. What I think you have in mind is rather the possibility of ethers of different *kinds*, each having its own specific function in the universe.

But this also is not new: it was a familiar concept in the eighteenth century, before our present knowledge of the interrelations of the different phenomena of nature was attained. Each of the various instances of apparent 'action-at-a-distance'—such as gravitation, the influence of the magnetic poles on iron, of electrified bodies on one another, of hot bodies on cold ones, of luminous bodies on dark ones—could conceivably be explained by the transmission of particles or through the medium of an ether in which the bodies concerned were embedded. There were many advocates of the latter idea, but since the phenomena were, so far as was then known, quite independent, each 'ether' had to have independent properties, so that the idea of multiplicity of ethers was quite common. They were not always called ethers—*fluids*, or *effluvia*, was a common term—but that is a mere matter of name.

It was not until the relations between the phenomena—first, the connection between magnetism and electricity, and then that between electromagnetism and heat and light radiation—were discovered that it was generally realized that a single ether was sufficient for all these things, and it was accordingly postulated. The idea of its uniqueness, therefore, is not an arbitrary assumption, but a reward earned by years of patient investigation. It is still not fully earned. An ether has not yet been conceived that would account satisfactorily for both gravitation and the other apparent action-at-a-distance phenomena, but it is not surprising that, with so much success behind them, physicists have a very strong belief that this is only a matter of time. Indeed, some have gone further and held that this same single ether is operative also in psychological 'actions-at-a-distance',[1] but I think very few are prepared to speculate so far ahead of our present knowledge.

When, in relativity theory, the ether is said to have been dispensed with, what is meant is really that this single ether has been deprived of its office as a standard with respect to which the motion of bodies can be observed and measured, and renamed a 'field'. The attempt to reach what would formerly have been described as a single ether is now known as the search for 'a unified field theory'. There have been proposals for such a theory, but the most that can be said for even the best of them is that it

---

[1] 'If any one thinks that the ether, with all its massiveness and energy, has probably no psychical significance, I find myself unable to agree with him.' Sir Oliver Lodge, *The Ether of Space*, p. 114 (Harper, 1909).

may or may not have succeeded in expressing some aspects of both sets of phenomena in a single set of mathematical terms, but it has not advanced our knowledge by suggesting any possible observation by which it may be tested.

There is therefore no conclusive reason why a number of ethers should not be postulated, each for its own realm of phenomena; and indeed, in such widely different realms as those of physics on the one hand and life and mind on the other, this may very well be the most practicable way of making progress. But within physics itself it would, I think, make very little appeal to scientists. Although the sufficiency of a single field for both electromagnetism and gravitation is, strictly speaking, still but a pious hope, I think its denial would be regarded by physicists as a very gross impiety and an abandonment of hope.

V

RETROSPECT AND PROSPECT

LORD SAMUEL:

1

You may have noticed that I have made no comments on several of your answers to points of mine with which you have disagreed; but that has not been because I consider them unimportant, nor yet because I can find no answers and must needs let them go by default. The reason is quite different. I have long felt that one of the main weaknesses of philosophy itself, all through its history, has been the tendency of philosophers of each generation in turn to devote their energies, first and foremost, to destructive criticism of the theories of the preceding generation: only afterwards, and often only perfunctorily, do they turn to the harder task of building up positive, constructive ideas of their own. No doubt the negative side is also essential. Francis Bacon quotes Antisthenes, 'being asked of one, what learning was most necessary for man's life? Answered, "To unlearn that which is nought".' No doubt there is much truth in that—and particularly during the last hundred years when most of the assumptions or postulates surviving from the mediaeval Schoolmen have been made obsolete by the revolutionary discoveries of modern science. In suggesting this Dialogue, and having by good fortune been able to enlist your co-operation, I have had the hope—which I am sure you have fully shared—that the book should be primarily constructive, and negative only when unavoidable. A constant watchfulness is essential, and a considerable part of our dialogue has come to consist of controversies between ourselves, but I have been anxious that that should not be predominant. I do not want the reader to feel, when he asks himself at the end, what does it all amount to? that he should be inclined to say that the book consists mostly of Herbert Dingle's disagreements with Herbert Samuel and Samuel's disagreements with Dingle, so that in the main they cancel each other out and he is left at the finish very much where he was at the beginning. That is far from being the case, but we ought to be careful that the book should not have that appearance. I hope you will not think it discourteous if, for that reason, I do not now elaborate my rejoinders to your replies and have kept them as short as I could.

2

The first of these points arises on our section on physics. I had put forward a proposition that there must exist in nature a universal continuum of some sort—some kind of ether—in order to make possible the transmission of light and heat from one place to another. I claimed that this is proved, in human or other experience, by the fact which is indisputable, that the light and heat from the Sun, for example, can be interrupted, wholly or partly, at any time and at any point, by the interposition of an opaque material object—by an eclipse of the Sun by the Moon for instance; or by a cloud being wafted across by the wind; or by our putting up our hand or opening an umbrella to shade our eyes.

You say in reply—if I do not misunderstand you—that you do not feel called upon to admit any such conclusion because you hold that all astronomical observations to be useful must be 'artificial', that is to say must be admitted to the record of human experience through scientific, including mathematical, procedure: but the movements of stellar bodies are beyond human control and therefore, you say, cannot be made the subjects of observation and experiment and scientific verification.

But why should we gratuitously hamper ourselves with any such limitation? If, for instance, I see a group of cows on a hot summer day moving across the meadow out of the rays of the Sun into the cool shade of the trees—is not that an observed physical phenomenon? Is it not as worthy of the attention of science and philosophy as anything the astronomer can produce in his observatory, or the physicist in his laboratory, or the mathematician sitting at his desk?

The marvellous achievements in recent years, first of the Russians and then of the Americans, in making artificial satellites and putting them into orbit round the Earth, have won the admiration of the whole world: the word 'sputnik' has come into the vocabularies. But I do not see why the observation of any layman, watching with the naked eye a solar eclipse, should be considered any less valid and significant. The Moon is nature's sputnik. And the natural satellite has advantages over the artificial one. It is always there: the other disappears or burns itself out after a comparatively brief career: the Chaldaean and Chinese astronomers recorded eclipses of the Sun hundreds or thousands of years ago, and we can predict others for next week or next year with confidence. Also the Moon is much larger than anything that our scientists are likely to put into orbit; so that, at certain times and seen from certain places, it fits nicely into the line of sight, and the whole disc of the Sun is eclipsed, enabling the corona to be studied and some Einsteinian predictions as well. Further, and not less important, nature's sputnik is provided gratis, and does not involve immense drafts upon the State Treasuries.

You add, however, this further observation, which is of a different kind: 'suppose', you say (p. 48), 'that, with the progress of knowledge, we find ourselves impelled to postulate that the solar system is an organism, in which a sort of super-mind bears the same relation to the movements of the planets as our minds to the circulation of our blood . . . it is possible that a time will come when what we shall call the "true" explanation of eclipses will be a psychological one, and the present account will become as trivial as a geometrical description of the path of the blood through the arteries and veins'. That no doubt is possible, but is such a possibility a reason why we should, meanwhile, in the present state of human knowledge, take refuge in an escape-clause of this kind? The fact that the intervention of a material opaque object, at any point at any time, in the line of sight between a human eye and the Sun, causes an interruption cannot be gainsaid. It is a proof that a physical process of some sort is going on continuously between the Sun and the Earth—and likewise in every other similar case; and this renders unnecessary the acceptance of so indefensible a conception as the figment of 'action at a distance'.

3

My second point relates to your insistence upon rejecting the proposition, which I have supported, that the subject matter of philosophy must be a 'nature' that is independent of human experience and is often different from it. You reassert your own position when you say (p. 138)[1]: 'The fundamental problem of all philosophy, including science, is, as I see it, to understand our experience, to see it as a rationally connected system in which each element is related to every other, instead of as a succession of apparently independent and unrelated happenings as it first presents itself. No matter whether we express our problem as that of understanding the universe, or discovering why things are as they are, or in any other way, it all comes down to the description I have given because all our knowledge of anything at all comes to us through experience—in the most general sense, of course, not sense experience alone.'

When I look for any reason why we should put this heavy responsibility upon one's own experience, together with the sum of other people's experiences, the answer that I find is only this (pp. 162-3): 'Experience is to me our *primary* datum, simply because it seems to me obvious that it is. Even if there is a universe behind it, it is only through our experience that we can know anything about it, and therefore our philosophy, if it is to be defensible, must in the last resort be based on that and must justify itself by appeal to that.' You proceed from there to say, in several passages in this dialogue, that nature and experience are equivalent concepts: 'nature i.e. experience'.

I do not know how this doctrine will appear to other people, but to me it seems that this foundation cannot support the vast philosophic structure you would build upon it, but would quickly crumble. To say that experience is to you a primary datum because it appears to you obvious that it is, would seem to me, to quote your own phrase, 'to reduce philosophy and science to a futile pursuit'. But to set out my reasons yet again here would be mere repetition.

4

I have only one other point: it relates to your use, on several occasions, of illustrations drawn from fictional literature as witnesses in favour of your view. No doubt a line of fine poetry or a noble piece of rhetorical prose will often lend point to an argument, besides giving aesthetic satisfaction of a high order; and

---

[1] See also pp. 114, 142.

imaginative fiction may provide an emotional stimulus to the philosopher and to the scientist: but I submit that they are not on that account entitled to be ranked as philosophic or scientific. Remember that wise saying of Dr Johnson: 'We may take Fancy for a companion, but we must follow Reason as our guide'.

When you would put Frankenstein and his 'monster' into the witness-box I would challenge his right to be there. You say (p. 36), 'Suppose we ask the question: who created Frankenstein's monster? There are two possible answers: (a) Frankenstein; (b) Mary Shelley. Both are correct within their respective contexts, but we must not mix them.' I would suggest that a third answer is possible—and would be the right one. No one created the monster because it never existed; the word 'creation' is therefore inadmissible. If it should be said that some day there may be constructed what we would now call a robot, but with an independent life and a personal mind, the answer would be that that might conceivably be so, but that it will be time enough for philosophy and science to take account of such a creature when it is brought into being: no ground has yet been given for supposing that that can be done.

Similarly with Prospero. You say that no one questions that Shakespeare created Prospero, and I would give the same answer. The essential characteristic of Prospero is that he is supposed to be a 'magician', with a book of incantations and his slave, a sprite 'Ariel', a being of a type described elsewhere as an 'airy nothing': these enable him to affect real events—to raise a tempest at sea and afterwards to allay it, or to bring about any kind of events outside the order of nature that he might wish. Such a personage never existed, and never could exist, so that it is inadmissible to speak of the 'creation' of Prospero. What is real is that Shakespeare wrote a stage-play, in which actors pretending to be living personages make believe to do such things as those; but none of it can claim a right to engage the attention of the philosopher or affect the conclusions of the scientist.

If literary or other fiction is to be brought in, how far are we to go? Are we to take seriously, for example, all the creatures 'created' by the delightful imagination of the Rev. Charles Dodgson of Christchurch College, Oxford, who, under the name of Lewis Carroll, was the author of the immortal *Alice in Wonderland* and *Alice Through the Looking-Glass*? Shall we try to classify zoologically the Mock-turtle, the White Rabbit,

Tweedledum and Tweedledee, packs of cards and sets of chessmen which have come to life and had their well-known adventures? If not, why not, if Frankenstein and Prospero are to be admitted?

You say that theoretical philosophy and science must take all possibilities into account, 'even if the probability is only of the order of one in a million'. But why? Why should we spend the limited energies of our brief lives in studying the abnormal rather than the normal? And what if the probabilities are of the order of nil in a million—as in the case I have just put?

There is another piece of fiction, which you also cite: it played a part in the controversy between the evolutionists and the fundamentalist theologians a hundred years ago—the attempt of Edmund Gosse's father (and, if I remember rightly, of some others also), 'to reconcile', as you state it, 'the findings of geology with the Scriptures by the postulate that the world was created at about the same time as man, with the fossils inserted ready-made in the rocks' (p. 35). You take this seriously and describe it as 'fascinating'. But forgive me for saying, I would rather call it preposterous, and a fraudulent attempt at the forgery of historical evidence which did little credit to its inventors, and even less to the moral character which they would ascribe to the divine Creator of the universe.

5

That ends my comments on your criticisms that have been left outstanding. I now revert to your positive ideas, to the basic principles which you have set out so fully and so clearly. I regret that here the disagreement between us is still unbridged, and is indeed emphasized by your latest contribution. In reaffirming that disagreement I must add a few words to explain how, as it seems to me, it has arisen.

Your central point is that you take the experience of each man and all men as the basis for your philosophy; and we must begin by analyzing this idea. You say that it does not require any other postulates than that the human self 'is alive and thinks'. But I would object at the outset that you cannot possibly stop there. The individual man can never be thus insulated: to be alive there must exist for him a physical environment. In the first place, he could not remain alive if he were denied an atmosphere, containing oxygen, to breathe: he would be dead in ten minutes from asphyxiation. Further, he must have some kind of habitat, some place in which to dwell and to move; and, as human beings are in fact constituted, that will be, and is, in fact, the solar planet which we name the Earth. Next, to be alive it is necessary first to have become alive: that is to say, being a mammal biologically, he must have had a father and mother: these must also have had fathers and mothers, together with their environments, and so on indefinitely. Again, having become alive he needs also to remain alive, and for that requires food and shelter. As things are, this too involves the existence and the activity of other people. As to thinking—he cannot do that in the ways that men do in fact think, without there being other like organisms to communicate with and to share his thoughts, through vision, speech, and in other ways. So that, having posited an individual who is alive and thinking as the foundation for your philosophy, you will be led on, step by step, until, as a consequence, the factual existence of such environments is also established. And these, put together, turn out to be nothing other than the Cosmos of which we have cognizance through our senses. In a word, a person, alive and thinking, implies a universe.

6

And now, turning to the positive side, we may ask ourselves—and the reader will certainly be asking—What is the upshot? What does it all amount to?

On my first proposition—the need for a continuous effort on the part of philosophy, science and religion to acquaint themselves with each other's activities and to bring into the forefront subjects on which they agree rather than those on which they are in conflict—we are wholly at one. And it is a principle that has undoubtedly been gaining ground in recent years, and ever more conspicuously at the present time. Even in the last few months, we may note many declarations by leading scientists and philosophers, with some from men of religion, and others from spokesmen of public opinion in general, couched in a very different tone from anything that could have been expected thirty or forty years ago.

One of the most weighty and most recent of these pronouncements comes from Sir Cyril Hinshelwood, chosen by the Royal Society to be its President in the tercentenary year of its foundation. As the main topic of his Presidential Address at the Anniversary Meeting on November 30th, 1959, he took the same subject as we have been discussing. I have asked his permission to quote several extracts, and this he has readily given.

Sir Cyril writes:

'Most men of science would disclaim any great concern with philosophy and especially with metaphysics . . . Yet in science the total neglect of philosophical aspects may be unfortunate. It can lead, on the one hand, to statements which are sensational but unsound, and, on the other, to an unhealthy indifference to the relationship of science to existence in general. And in any event, even though science may seem to decline metaphysical statements, it not only throws out challenges to philosophy but is frequently itself confronted with problems inescapably philosophical in nature.

'Scientific theories, though less ambitious than the systems of the older philosophers, and though claiming neither universality nor finality, do aim to be comprehensive and do hope to stand the test of time as long as possible. Moreover, we do judge them by a criterion in which it would be hard to disclaim every metaphysical tinge, namely, by the degree of mental and aesthetic satisfaction they afford. However much we may protest, utility is not enough. There is undeniably, in the law of gravitation some quality distinguishing it from a set of empirical formulae, however convenient. Men of science do, unconsciously at least, place hypotheses about nature in a sort of hierarchy of esteem, where the conservation of energy and the second law of thermodynamics rank high and mere working rules rank low.

'The difficulty of complete aloofness from philosophy is illustrated by parts of physics . . . Everyone, whether he admits it or not, seeks something which he calls understanding. The professed empiricism about wave mechanics is half-hearted, and more often than not the abstract mathematical formulation is given a metaphorical translation in which symbol and reality become very much confused. "Waves of probability" are created, electrons become "clouds", potential barriers are "tunnelled through" . . . Incidentally it may be remarked that if the more serious and responsible men of science remain aloof from the task of relating their subject to its philosophical background, the less serious and responsible will do it for them, and will startle the uninitiated with fantastic paradoxes which may impress the layman, but do science real harm . . .

'Nor are the discovered facts of nature merely to be made the basis of formulae for the sole purpose of computations. In the last resort what men of science search for is the vision of nature, and this must be expressed and communicated in those terms which provide the nearest approach to real understanding.

'If this is true of the physical sciences it is equally so of biology. One of the transitions from physics to biology is illuminated by a remark of Heisenberg, "The mathematical formulae indeed no longer portray nature but rather our knowledge of nature".'

7

I propose to add later, in other contexts, some further extracts from Sir Cyril Hinshelwood's timely and cogent Address. But I would mention now, as another example of the present trend, some observations by the Master of Balliol in his annual Report on the College for 1959. They are specially significant coming from him because Balliol has for many years been in the forefront among the Oxford Colleges in specializing in the study and teaching of Literae Humaniores. Dealing with the number of undergraduates now in college, which is too large to be handled efficiently, Sir David Keir says 'It becomes an interesting problem how to achieve at a lower figure a balance that will satisfy three conditions: first, recognition of the growing importance of science in modern life and education at all levels; second, maintenance of the humanist tradition, *of which professional scientists and technologists themselves are now numbered among the strongest defenders*'. (I have underlined an observation which seems just now of special significance.)

I feel sure that many similar opinions could be cited from leading educationalists, not only in this country, but in many of the others in which thought is free. I would quote one example.

The Massachusetts Institute of Technology is generally regarded as the leading institution of its kind in the United States. On my last visit to America, in 1952, I went to Boston mainly in order to see the MIT at work, and especially to get information on this very point—how to prevent, in education at the university level, the humanities being overwhelmed and submerged by technology. Our Parliamentary and Scientific Committee, consisting of Members of the House of Commons and House of Lords, with representatives of the chief professional organizations of the applied sciences, with which I had been closely associated for many years, had been much concerned about this problem; they had put it in the forefront of their inquiries and activities. The heads of the MIT were most helpful in answering my questions, and I left heavily documented with reports on the curriculum. I found that they were as much concerned about this danger as we were. They were already insisting that all their thousands of students should give a large proportion of their time to the humanities—no less in fact than twenty per cent, spread over their four-year minimum course; they included even such subjects as music and modern languages.

Across the River Charles, a few miles away stands Harvard University, and they too, I was told, were anxious about this matter, approaching it from the opposite angle. The whole situation was summed up in a saying current in both establishments —'It is the business of MIT to humanize the scientists, and it is the business of Harvard to "scientize" the humanists'.

I would include in this short anthology one more testimony— the Christmas Message for 1959 broadcast to the Commonwealth by H.M. the Queen. It ended with these sentences: 'In recent years the Commonwealth countries have been making a great co-operative effort to raise standards of living. Even so, the pace of our everyday life has been such that there has hardly been enough time to enjoy the things which appeal to men's minds, and which make life a full experience. After all, our standard of living has a spiritual as well as a material aspect. The genius of scientists, inventors, and engineers can make life more comfortable and prosperous. But throughout history the spiritual and intellectual aspirations of mankind have been inspired by prophets and dreamers, philosophers, men of ideas and poets, artists in paint, sculpture and music—the whole company who challenge and encourage or entertain and give pleasure.'

I wonder whether, in modern times, any other Royal Message had been bold enough to make mention of 'philosophers and men of ideas'.

8

I had written this brief outline of various features in present-day physics which seem to be open to criticism—but always with the doubt at the back of my mind whether it is not an impertinence for any layman to venture upon such interference—when my attention was drawn to the Annual Report of the Advisory Council on Scientific Policy.[1] One paragraph has a direct bearing upon the point I am discussing. The paragraph was this.

'School Science Curricula.

'46. We have no doubt that school science curricula are in need of a thorough re-examination. They tend at present to be unimaginative and to be overloaded with factual material (in part as a result of the tendency to keep adding new material without removing the old). It has been suggested to us that up to twenty to twenty-five per cent of the content of the curricula in physics, chemistry and biology could be removed without any harm—and indeed with benefit. Mathematics curricula are equally in need of revision.'

This Committee, appointed by the Government and consisting of a number of leading figures from all the principal branches of British science, is authoritative. If they say that our present curricula in physics, chemistry, biology and mathematics are in need of a thorough overhaul, and have suggested that up to twenty or twenty-five per cent of their content are ripe for discard—for that is what it comes to—an interloper, such as myself, may take courage. The most complacent of present scientists can hardly ignore such a pronouncement from such a quarter.

Lord Adrian adds the weight of his authority. In a debate in the House of Lords on the need for the training of more scientists, he said[2]: 'The curricula in science can certainly do with a thorough re-examination, as the Council believe'.

9

When we turn to the other branch of our proposition—the association of philosophy and science, not only with one another but of both with religion, we see some signs today of the same trend. This is so mainly in the western world, but it is to be seen in

---

[1] Cmd. 1167, 1960, published in October 1960.
[2] House of Lords, November 1960.

some countries of the East, particularly in India and in Turkey. In this sphere the obstacles are even greater, for all the theologies must come into the field of discussion. But it is now a hundred years since the shock of controversy was felt in England when the intellectual world was split in two by Darwin's theories, which were in flat contradiction with the cosmology that was then held to be an integral part of Judeo-Christian religion. The clash culminated in the famous debate at the meeting of the British Association for the Advancement of Science at Oxford in 1860, when the champions were T. H. Huxley for Darwinism and Bishop Wilberforce for the Book of Genesis. Born in 1870, I belonged to the first generation after the dust of that historic battle had cleared away. Indeed, I had the privilege of hearing, in the Sheldonian Theatre at Oxford, Huxley himself deliver the last, and best known, of his lectures—the Romanes Lecture of 1893, on *Ethics and Evolution*. A bent old man, a pathetic figure, looking very small in the centre of the floor of Wren's spacious building—no microphones and loudspeakers then—he was hard to hear in the undergraduates' gallery. The coming generation of that day, we were almost all whole-hearted Darwinians. I think it would be the same with the youth of today.

Let me cite again the tercentenary, in the present year (1960), of the Royal Society. Among the principal functions has been a special Service attended by the Society in St Paul's Cathedral, the sermon delivered by the Dean, Dr W. R. Matthews. It was natural for him to recall that controversy of just a century before, and to compare the situation today of religion in relation to science with what it had been then.

The sermon was a remarkable address, forthright, cogent, the work of an eminent thinker, of a mind at once philosophic and devout. It has not yet been published, but the brief reports in the Press made it clear that it was completely in tune with what we have been saying in this Dialogue, and that it would greatly strengthen our case if we could repeat some of its salient passages. A friendship of long standing permitted me to ask Dr Matthews if he would allow me to quote extracts from the script from which he had spoken. To this he readily agreed, and I end gratefully this part of my contribution by supplementing my quotations from the President of the Royal Society from the standpoint of science, with the following from the Dean of St Paul's from the standpoint of religion.

Referring to the controversies on evolution of a century ago the Dean said:

'Very few educated Christians today could sympathize with the stand taken then by the representatives of orthodox Christianity. They exhibited a stubborn resistance to new knowledge and a reluctance to examine evidence which we can scarcely understand. Yet these men were intelligent and honest, some of them, like Gladstone and Wilberforce himself, forward-looking and reforming in other spheres. What then was the cause of their violent reaction against the new hypotheses? They believed that they were defending a truth so precious and so fundamental that any apparent attack upon it, or weakening of its authority over men's minds, must be repelled. In my opinion, fundamentally they were right. The belief in God the Creator and His revelation held a truth that mankind cannot abandon—it is not only a truth, but a saving truth. But they were wrong, disastrously wrong, in thinking that this truth depended upon the literal accuracy of the Creation Myths in Genesis. They were wrong too in thinking that the Bible was a source of scientific knowledge. And this error led them to a further error which really contradicted their own faith. They believed that God was the eternal wisdom and truth and, therefore, that all truth is in some degree a revelation of God. They ought, surely to have concluded that no truth could be unholy and all claims that new truth has been discovered ought to be candidly examined . . . Neither Science nor Religion has spoken its last word. It is certain that before the next centenary of the Royal Society comes round all the current text-books will long have been obsolete. Religious thought moves more slowly, but it moves, partly under the stimulus of science. I pray that the course of development in both may bring them closer together, each learning from the other, so that these two majestic forms of the spirit of man may unite in the worship and service of the God of truth.'

Is not this in close accord with the pronouncement of Huxley himself, in his *Lay Sermons*—'Learn what is true in order to do what is right, is the summing-up of the whole duty of men, for all who are unable to satisfy their mental hunger with the east wind of authority'.

10

Writing at a later date, January 1961, I would add a note which will illustrate the existence nowadays of a trend towards a closer association between theology and science and philosophy. The University of Birmingham has been advertizing a vacancy in its Chair of Theology, and it does so in the following terms:

'The University of Birmingham is undertaking a development in its Department of Theology of more than ordinary interest. A second Chair, provided by the Edward Cadbury Trust, has been established for a period of seven years in the first instance. This Chair will be filled by someone whose interests stretch across the borders between the natural sciences, philosophy and theology. It is hoped that the new professor will pioneer research and teaching in this general area.

'In Birmingham the sciences have always been strongly represented, and, at present, half the students in the University are reading for degrees in the Faculty of Science. Moreover, last October the Faculties of Arts and Law moved to Edgbaston, and, for the first time for nearly sixty years, all the departments of the University were gathered in a single place. This development seemed to offer an opportunity for interesting and profitable work in exploring the little known territories where theology, the natural sciences and philosophy meet. The Department of Theology itself, now twenty years old and growing steadily, is non-denominational and widely representative.'

I wrote to the University asking for a copy of the advertisement, and the Registrar, Dr Geoffrey Templeman, when sending it, was good enough to add this note:

'The background to the present situation is this. We, like many of the modern universities, have only instituted the study of theology comparatively lately. Our own Department has hitherto confined the sphere of its teaching and research within the general field of the humanities. On the other hand, this University, as perhaps you know, has always been heavily weighted towards the sciences and medicine. At present rather over half of our 4,600 full-time students are registered in the Faculty of Science . . . We hope to find someone for this Chair who is a scholar able to promote research and teaching in the borderland where theology, the sciences and philosophy meet.

The unusual form of the advertisement is explained by the fact that we know the task of finding a suitable candidate is going to be very difficult, and at this stage the Electoral Board is seeking help and suggestions publicly and privately.'

There is no reason to suppose that the present atmosphere in Birmingham is any different from that in other modern universities in this country; and this example may serve as an answer to any who may object that nothing is so unchangeable as theological dogma, and that our enterprise here is quite unrealistic and hopeless from the beginning.

And now there comes another example of the same trend, from a quarter even more authoritative than any of these. The Convocation of Canterbury of the Church of England passed, unanimously, on January 18th, 1961, the following motion: 'That this Convocation, recognizing the vast range of modern scientific discovery, welcomes the increasing study being given to the moral and spiritual significance of these discoveries for human life and conduct and encourages further study'. While the retiring Archbishop of Canterbury, Dr Fisher, on March 8th, 1961, speaking as president of the Society for the Promotion of Christian Knowledge (*The Times*, March 9) said: 'There was a great theological reformation running through the whole of Christendom. It would be quite wrong to think that the theology, even of the Roman Catholic Church, was stable. It was changing, too, and would change a great deal more in the future.'

11

I revert to my text. That satisfaction with our agreement so far has unfortunately not lasted long.

I turn to my second proposition—that philosophy and science are most likely to reach definite results if each starts by dividing its sphere of study into two fields: one, the cosmos as it is in itself, given to us by nature; the other, the universe as it is for us, the world as it is apprehended and portrayed by Man. But when I put forward that proposition, purely as a matter of procedure in discussion, you demurred at once and expressed a disagreement which you described as fundamental (p. 34). All through our debate that divergence has continued, reappearing again and again in various contexts. Your objection is not to the validity of any of the particular examples that I have given in support of my thesis; nor do you deny that there may exist a cosmos 'as it is in itself': you say indeed that we may 'end by discovering it'. Your objection is *a priori*: you say that human philosophers can only have access to human experience, and that 'our ultimate aim is to unite all experiences under a single set of laws'.

You say again (p. 114), 'To me, just as there is no wealth but life, so there is no reality but experience ... Perhaps, to attempt the impossible ... I ought to say that experience is the only *objective* reality, using "objective" in the grammatical sense and not as the equivalent of "external".' You speak again (p. 142) of 'Nature i.e. experience'. And once more (p. 193) you say: 'My equivalent of your "nature" is experience—that which is presented to me for study. It seems to me the inescapable starting-point of philosophy, for even if you postulate an external universe, all that we can discuss concerning it is the aspect which it presents to our experience; anything outside that must necessarily be inaccessible to us.'

But I remain unconvinced.

I do not think you have substantiated that principle, and do not believe that it is possible to do so. Rather I would submit that the opposite is true, and that nature is *not* to be identified with experience: it includes mankind, and human experience with it, but it includes much else as well.

Before I embark on a final review of this argument I would point out that the word 'nature', in common parlance, is itself a source of confusion. It is frequently used as denoting the Universe without Man. We speak of 'Man confronted by the forces of Nature'; of Man, with a higher morality, establishing civilizations, and engaging in an unceasing, and often desperate struggle to counter the evils that nature brings upon him (the other, and larger part being the sufferings and disasters that man brings upon himself through his own mistakes, follies, crimes, vices and sins). But it is surely the business of philosophers to recognize and, if possible, eliminate, this ambiguity. It is a source of continuous barren argument, for it leads to the treating of a purely artificial dichotomy as though it were a matter of substance and represented the situation as a whole.

I do not propose to repeat now the arguments which I put forward when you first raised this issue, but to turn to other writers for endorsement or criticism of the thesis.

12

An eminent British physicist who was interested in my previous book, *In Search of Reality*—which had been introduced to him by the late Sir Henry Tizard—was good enough to send me some helpful comments. On the question of an ether he writes: 'I think most physicists would agree that opinion is tending towards it. In general I would favour Lord Samuel's theory, but with the reservation that a theory of the aether expressed in words seems too vague to be much use and I don't see how you would put it into symbols.' I feel obliged to offer a reply to this objection, although it involves a brief recapitulation of points that have found a place in our previous section.

I must in the first place admit that evidently I had not sufficiently made it clear that, in the 'two-state ether', the active ether is conceived as something different from the quiescent—the same 'stuff' but differentiated by the essential that it is active; then it is capable of producing phenomena. I agree that my critic's objection is valid, but it is based on a misunderstanding: it applied to quiescent ether, but not to active.

The world of active energy would in fact be substantially the same as the present subject-matter of physics, including astronomy and chemistry. An acceptance of an ether theory of this kind would leave the work and the methods of the scientific laboratory mostly unchanged. Why, then, trouble about 'quiescent energy' at all? The reason is that it is the indispensable matrix of active energy: and it is that which gives to the ether its status in the universe.

You say that a universe as it is in itself apart from man is inaccessible to him. I would contend that it is not merely accessible—mainly perhaps by the road of legitimate inference—but that we cannot escape it. It is with him every moment of his life. If we have found it so difficult to discover, it is precisely because it exists everywhere and always, inside us and outside.

13

As to the reality and accessibility of a given cosmos, Max Planck is one of the eminent scientists who feels no doubt. He says, 'The laws of Physics have no consideration for the human senses; they depend upon facts, and not upon the obviousness of facts'.[1] Emerson said much the same—'Nature, which made the

---

[1] Max Planck—*The Universe in the Light of Modern Physics*, p. 69 (Allen & Unwin).

mason, made the house'—that is to say, man and all his works is the product of nature, and not the other way about.

Even Shakespeare, who sees so much more than other people, presented the whole situation in a few lines in *As You Like It*. In Act III, Scene 2, Rosalind, who has taken refuge in the deposed Duke's camp in the Forest, there meets Orlando: they exchange Elizabethan 'conceits' about Time. She addresses him:

'Do you hear, forester?'

Orlando: 'Very well; what would you?'

Rosalind: 'I pray you, what is't o'clock?'

Orlando: 'You should ask me, what time o'day: there's no clock in the forest.'

There we have the dividing line between the man-centred universe, with its arbitrary time intervals of hours, which might just as well have been conventionally twenty in a day, or ten, as twenty-four; and the Solar System with its revolving Earth, giving us the alternation of night and day. This is no more the product of human experience than is the existence of an incandescent Sun, radiating light and heat, with a planet at a distance, with physical conditions that make it habitable by organisms such as ourselves. Now in a scientific age, we are getting so accustomed to this distinction between the given universe of nature, and the man-centred world of human experience and activity, that the colloquial language of industry and commerce has to borrow terms from the vocabulary of philosophy, and we see in the newspapers displayed advertisements of 'man-made fibres': a century or two ago such words would have been regarded as impious, if not incomprehensible.

14

Let me now revert to some of the examples which would answer the inquiry: What would be 'the use' of an ether hypothesis to theoretical and applied science?

The null result of the Michelson-Morley experiment proved that the earlier conception of an ether, which regarded matter as one thing and an ether as another and separable, was wrong. But the conclusion that therefore no ether can exist, may also be wrong. A two-state ether, either quiescent or active, is not open to that objection. We are not inhibited therefore from discarding several conceptions which had been adopted to fill the gap supposed to have been left by Michelson-Morley. Among these are: (a) The notion that the transmission of light and heat, and other electromagnetic phenomena, can be effected by 'actions at a distance' and without any kind of process going on in between—which Newton called 'a great absurdity' and Einstein a 'phantom'. (Scientists, incidentally, would no longer feel obliged to acquiesce in the general use of such a nonsensical expression as broadcasting programmes being 'on the air'.)

(b) Nuclear research has discovered that in certain experiments new electronic particles may be produced, and when it is asked where do they come from the answer that is given is that 'they are created out of nothing'. Quiescent ether would provide a matrix out of which they could originate, and into which, if unstable, they would disappear.

(c) We could dispense with the figment of the so-called 'laws of Chance', which Professor Max Born (already quoted) said 'Certain observable events obey': he declared that 'Today in physics, chance has become the primary notion, mechanics an expression of its quantitative laws, and the overwhelming evidence of causality with all its attributes in the realm of ordinary experience is satisfactorily explained by the statistical laws of large numbers'. This, he said, 'is now generally accepted by physicists all over the world, with a few exceptions'.

(d) It will be possible to discard also the relic of scholasticism that lingers on from the Middle Ages, which requires us to postulate that material objects possess 'properties' or 'qualities', and that these are the causes of the phenomena in which they are engaged. You gave an illustration of a piece of lead whose nature could not be analyzed until those ideas had been abandoned. I followed this up by listing six different processes in which the lead participates. An ether would provide what J. J. Thomson called 'the seat' of those phenomena.

(e) In particular, the idea, which is still taught today in the science classes in the schools of London, that a brick, for example, which someone has lifted on to a table becomes different from a similar brick left on the floor, because it has 'acquired potential energy of position' by transfer of energy from the muscles of the person who had lifted the brick, and this energy can be 'released' and 'become kinetic' when someone moves it to the floor again—all this can be discarded, with the other 'properties', and we will be free to recognize—what should have been obvious long ago—

that a collection of molecules, such as a brick, cannot be found by observation and experiment to exhibit any such difference because of its past history and present situation. The phenomena are due, quite simply, to the Earth's gravity; the object's participation being an entirely passive response to the gravity pull. An ether theory would provide the medium for its transmission.

(f) It is no answer to say that, unless this actually happens, the law of the conservation of energy would not hold good. That no doubt is so. But it does not follow that an ether hypothesis must be wrong. It may be that it is the universality of the law of conservation that is at fault and may be brought into question. Many phenomena are processes which have a beginning and an end. As a rule, energy may be transformed into various patterns while its quantitative measurement remains the same. But this need not make us assume that that must be so always. There may be a class of cases in which a process of ether activation has been started and afterwards ended. When we light a match and from it a candle we set going first one process of incandescence and afterwards another. If we blow them out, each process comes to an end. We need not assume that the match and the candle have been able to draw upon a universal and eternal stock of physical energy, a part of which has here been 'conserved' and afterwards 'released' in equal quantity, and then absorbed—all this is only an assumption which, in particular cases, may not be justified by the facts.

(g) Einstein having established the non-existence of absolute motion, an ether theory might supply an alternative for the concepts of space, of time and of motion, which are the foundation for the paradox of space-juvenescence which you so rightly hold in contempt.

I would have wished to add two other examples of the possible value of an ether theory. One would offer an alternative to the theory of an expanding universe as the explanation of the so-called red-shift in the spectra of distant galaxies. On this I have given at some length, in an earlier section, reasons for a reconsideration. The other example would offer a different line of approach to the obdurate wave-particle problem. But both of these are clearly matters for the professional physicists, where the layman—philosopher or other—ought not to venture to do more than invite their attention to the possibility of a different approach leading perhaps to different conclusions.

To ask what would be the use of an ether hypothesis to theoretical or practical science is a perfectly legitimate question, which cannot be left unanswered; but it is not the decisive question. To enlarge the bounds of human knowledge is of more importance than any of those considerations that we have suggested in answer. To establish new truths is an end in itself. Whether any practical results can be seen beforehand—whether, that is, pure research 'will pay'—is not what matters most. If the taxpayers have to be persuaded, they can be reminded of the many cases in the history of science where disinterested research has resulted in developments in technology that were quite unforeseen, but which have become the foundations for new and prosperous industries of great economic value to mankind. Occasionally such discoveries may have been initiated by some clumsy suggestion from a scientifically-minded layman, intruding where he had no right to be.

Lastly—and most important of all—an ether, if recognized as a reality in the given universe, would allow a long step forward and a great simplification in cosmology. It would constitute a physical and philosophical basis for the universe; it would give to the Cosmos its unity in extension and duration—in common parlance, its unity in space and time.

15

The year 1960 saw the publication of two important addresses by leaders of British science which gave an authoritative account of the present trend of scientific thought on many of the outstanding problems. They touch upon several of the matters with which we have been dealing in this Dialogue. One is the Presidential Address to the British Association for the Advancement of Science, at its meeting at Cardiff, entitled 'The Two Aspects of Science' and delivered by Sir George Thomson, F.R.S.[1] The other is the Presidential Address of Sir Cyril Hinshelwood, O.M.,[2] on the occasion of the tercentenary of the Royal Society, from which I have already quoted. I propose now to include here extracts from both. If any of our readers are inclined to say that I am quoting too much from other people, let me explain that I do this because I am always conscious of my lack of the formal qualifications, academic, professional or technological, and am

---

[1] Sir George Thomson, F.R.S., Master of Corpus Christi College, Cambridge.
[2] Sir Cyril Hinshelwood, O.M., F.R.S., Professor of Chemistry at Oxford.

obliged to enlist the aid of those who possess those qualifications in full measure, and are able to command for their opinions an attention which I cannot expect for my own.

Also, when I try to practise what I preach and approach the unsolved problems from the philosophical and the scientific sides simultaneously, and when I find the President of the Royal Society and the President of the British Association proceeding on just the same lines, it is natural that I should take refuge under their authority. As to the association of both philosophy and science with religion, I can do no more than express the conviction that it is only if all the great religions will make the effort to free themselves from myth and magic, and embrace science and philosophy instead, will the world move faster than now in its progress towards individual good living and national righteousness.

On the question of the association of science with philosophy and religion, Sir George Thomson says:

'Science aims at understanding the nature of things; in this it is at one with religion and philosophy. But its approach is the opposite. These last try to gain knowledge of the whole, in the one case by an awareness of the deity, intuitive or revealed, in the other by building with words a system of thought which can account for fundamentals. Science starts from the other end. It begins by studying details, often apparently trivial details but things which are queer and appeal to human curiosity.' He says also: 'What we want as scientists—I am sure in this I speak for the great majority—is that the world should realise that we are not interested merely in making possible new drugs, television sets, or weapons, though all these are important, but in enlarging the bounds of human knowledge . . . To me science without technology is incomplete and inconclusive. Systems of philosophy come and go, some are perhaps true but who can tell? But when conclusions deduced from precise experiments by mathematical theory lead to detailed predictions from which working machines can be designed, machines which without the theory no one would have thought of in a million years, then indeed one knows that one lives in a universe which is rational and that one has found the key to one of its rooms.'

With regard to what I have called 'the two worlds', Sir George writes: 'Concepts are discoveries as well as, indeed more than, inventions ... Not merely are they an exercise of the human mind which equals the brilliance of any system of thought, philosophical or even mathematical, but they represent reality. The nature of the relationship is to me a mystery, but they are certainly not merely the product of the human mind ... As science advances, concepts tend to become more and more abstract, further from anything that can literally be touched or handled ... In the higher flights of theoretical physics abstraction goes much further. Is there not a danger that one may lose touch with reality, and end up by supposing that some elaborate piece of mathematics represents reality when it is only a creation of the mind, inspired indeed by physical reality but no more like it than is a modern picture?' And he concludes by saying: 'Science is not merely the control but also the understanding of nature. Its two aspects must be held in equal honour.'

Sir Cyril Hinshelwood writes to the same effect. 'The question of the internal and the external worlds,' he says, 'cannot and should not be ignored by men of science.' He singles out for mention 'one or two rather specialized examples'. He chooses these, he says, 'because they not only involve fascinating discoveries but because they lead us to ponder once again on deep questions which should not be ignored by the contemporary man of science. They relate, in fact, to the concomitance of the inner world of conscious experience and the outer world of nature.'

'Concomitance' is a key-word, and a very helpful concept. In effect, Hinshelwood recognizes the existence for philosophy of two worlds—the given Cosmos as it is in itself, and the man-centred universe as it is for us. We can see that both are evolving —the first, according to our standards, very slowly, the second sometimes very fast—and never faster than now. In the Stone Ages the difference between the two was relatively great; at the present time it is much smaller. The whole age-long effort of both philosophy and science is to bring about the 'concomitance' of the one with the other.

All this must apply to biology as well as to physics and chemistry, and so we come back to the question of life and mind. Here Hinshelwood's position is forthright. He gives no support to your argument that the mind-body problem is misconceived, and that, rightly viewed, there is nothing to be discussed. He says: 'The central problem of mind and matter is not always thought worthy of their attention by men of science, yet it was given profound consideration several decades ago by Sir Charles Sherrington, and we are vividly reminded of it in the most recent book of one of our senior Fellows, Lord Russell'. He proceeds later: 'The mind-matter relation is quite often dismissed as non-existent or meaningless. But moral qualities are not explained by natural selection, human behaviour is not machine-like and the mind-matter relation cannot be ignored in an intelligent consideration of existence.'

I have now nothing fresh to add to what I said about this in our preceding Section. The difference between the chestnut and the pebble, and between what birds can do and what trees can do, is a fundamental fact in a given universe. To bring in human experience is to engage in an interesting intellectual exercise, but the events and processes go on quite independently of what we humans think about them, or what our progenitors may have thought, or not thought, thousands or maybe millions of years ago. The philosopher who seeks to evade these problems does so at his peril. It is not enough merely to say, 'the actual problems, of course, remain': although we are obliged to admit that, we must wait for further discoveries from the devoted labours of the keen biophysicists and biochemists before we can hope to do better than Sherrington's duality.

PROFESSOR DINGLE:

1

Our discussion could indeed continue, and unless we stifle its immortal longings and bring it to a violent end, there seems no reason why it should cease. I am wholly with you in hoping that it shall be constructive, and I do not think it need be the less so if, at the end, we must agree to differ on fundamental points. We have set out our views, and our reasons for holding them: if the reader is thereby helped to clarify his own thinking, I think something valuable will have been achieved. I will therefore make only such brief comments on your final remarks concerning our disagreements as seem necessary to protect my position from misinterpretation.

2

In the first place, I do not say that only artificial occurrences can 'be made the subjects of observation and experiment and scientific verification', nor do I hold that we should 'gratuitously hamper ourselves with any such limitation'. It is a matter only of practical procedure, not of fundamental principle. All experience is our subject-matter, but the discovery of science is the discovery that we can make artificial experiences (by arranging the conditions under which we observe—in short, by making experiments) which show simple relations with one another, whereas the things that happen naturally are baffling in their complexity.

Let a piece of paper fall from your hand. Repeat the operation twenty times, as nearly identically as you can. You will be incredibly lucky if the paper comes to rest in the same position twice. You can get no law of falling bodies from that. But do it in a vacuum, with a standardized mechanism for releasing the paper in the same manner from the same place every time, and you will then be able to make observations that will help you to find such a law. When you have got it you can apply it to some natural movements—e.g. the movement of the Moon—with considerable, though not quite perfect, success. This would not have been possible without the artificial experiment.

You know Browning's lines:

> That low man seeks a little thing to do,
>   Sees it and does it:
> This high man, with a great thing to pursue,
>   Dies ere he knows it.
> That low man goes on adding one to one,
>   His hundred's soon hit:
> This high man, aiming at a million,
>   Misses an unit.

Browning, of course, is extolling the 'high' man because he rates aim higher than achievement. His 'high' man is the pre-scientific philosopher; his 'low' man is the scientist. But what he does not tell us is that the 'low' man may have the same aim as the 'high' man, and that by adding one to one he will eventually reach not only a hundred but the 'high' man's million, the 'high' man in the meantime having missed the unit. The 'high' man is impatient and arrogant, the 'low' man patient and humble. The temporary restriction to artificial experiences does not 'hamper' us with a 'limitation'; it is a practical device for solving difficulties invulnerable to direct attack.

But, having said that, I cannot help expressing admiration of the way in which you consistently keep the ultimate *end* in view, unlike the bulk of our mathematical physicists, who are open to Browning's criticism; they are so enraptured by the temporary *means* that they have lost the power of perceiving the end. You have not forgotten that the final aim of science is to understand such things as cows moving towards trees, while most of your critics have never conceived it to be other than the construction of a differential equation—a Supreme First Equation, of course, whose solution is the universe of their fashioning; but still, an equation. The discovery that we can make artificial experiences which we can measure and fit into mathematical theories has opened the door to such a wealth of possible activity that it has been almost universally overlooked that this is but a device. When, if ever, the true end is reached, mathematics can fade away so far as science is concerned, because any measure-numbers that we could get by applying our instruments to what happens naturally would not only be intractably irregular, but also would not be of any deep significance because they would inevitably denote not natural experience itself but the response of something natural to our artificial instruments. Mathematical physics is a means—an indispensable means, so far as we can see—of understanding nature (or experience, according to one's point of view), but it is only a means.

But that our mathematicians have become so bewitched by it that they have come automatically to regard it as an end in itself, is exemplified clearly enough in the remark of Bondi's which I quoted earlier (p. 143): 'One of the weightiest objections brought against the hypothesis . . . is that it has not proved possible, so far, to put it into a mathematical theory'. It does not matter what this hypothesis is, and I am making no comment on it; I am considering only the attitude unconsciously exposed by this remark. If, of any hypothesis at all, it can truly be said that one of its greatest defects is that it cannot be put into a mathematical theory, we should hail it with rapturous delight: of how few hypotheses can it be said that they have no greater defects than this! It might have been inconsistent with fact, too narrow in scope, far-fetched, purely *ad hoc,* or afflicted with any other of the substantial ills to which hypotheses are prone, but it has done nothing worse than reach its goal without deviating into mathematics. Yet its outstanding merit is described as a 'weighty objection'. It is as though an economist had become so obsessed with the normal means which we have to adopt for making generally available the necessities of life that he could say, of a recommendation that dwelling-houses should be well ventilated: 'One of the weightiest objections to this proposal is that it has not proved possible, so far, to manufacture fresh air and distribute it at a price within the reach of all'.

I am anxious, however, not to give the impression, by selecting particular examples, that I am selecting particular persons for criticism. I particularize in order to avoid vagueness and the suggestion of speciousness, unavoidable in a merely general statement, and in doing so I am careful to quote what is *typical*, and not peculiar to the particular instance chosen. Bondi's remark reveals so clearly the unconscious general attitude that it suits my purpose here, but it is no worse than that of the Royal Society referee, for instance (p. 72), who dismissed what was not mathematical as 'a smother of words', without troubling to consider what the words meant, or than Born's dismissal of the plain meaning of a word by describing it as a 'force of suggestion' (p. 80), or than a host of other cases which are immediately obvious to anyone who reads modern mathematical physics with unclouded eyes focused on the goal.

3

On the matter of illustrations from fiction, your statement that 'No one created the monster because it never existed' presupposes the particular meaning of 'existed' which you postulate, whereas we do, in fact, use the word equally validly in other senses, with correspondingly different meanings for 'created'. When you consider Shakespeare, Prospero and the spirits, you are dealing with three objects of discussion which are essentially different. In the framework of your 'realistic' philosophy you can only say that Shakespeare 'existed' while Prospero and the spirits didn't, but that takes no account of the fact that there is an 'existential' difference between Prospero and the spirits which is essential to the play. I can express that difference by saying that, within the framework of the play, Prospero exists, while Shakespeare is a meaningless word and the spirits are phantoms; but in the framework of ordinary commonsense 'realism', Shakespeare exists and creates the concept of Prospero creating the illusion of spirits. You want to claim a proprietary right to the term 'existence' for commonsense realism, but that would not only transcend the conventions of ordinary speech, but would also rule out of consideration possibilities which may very well be realized.

For example, there are those (Carlyle was one) who take the speech, which Shakespeare puts into the mouth of Prospero and which you quote on p. 176, to describe an actual possibility in your 'realistic' sense, and believe that the great globe itself—this very tangible, real world of ours, that 'existed' before there was any life at all—will dissolve, exactly like the 'insubstantial pageant' of spirits, and 'leave not a rack behind'. I express no opinion on that, but I must point out that, since this is conceivable, it may, so far as our knowledge goes, actually occur (speaking in your terms: in mine, it may be the only available rational interpretation of experience). One very commonplace way would be that you might wake up and find that Shakespeare, the globe, the stellar universe, and all the rest were merely parts of a dream, and that 'reality' is something totally different from all this.

You start with a postulate that would make this impossible: these things 'exist', and there's an end on't. Such a starting-point dogmatically prohibits what reputable thinkers believe to be not only possible but actually to be realized. I am not content to do that. I want to start at a position from which everything conceivable is a possibility, and then narrow down the possibilities into a much smaller field of actuality by examining what experiences do in fact occur and discovering how they are related with one another. I can then assign appropriate status to Shakespeare, Prospero and the spirits; set them in consistent relations with one another and with the rest of experience; and then proceed to investigate those relations in more and more detail.

But I need not repeat what I have already said on this point. My object now is merely to clarify my position, not to defend it, and I want to make it clear that I believe that the imaginings of Shakespeare have just as much 'right to engage the attention of the philosopher or affect the conclusions of the scientist' as the motion of the Moon has. But the scientist will not classify Shakespeare and Prospero together as similar beings, any more than he will 'classify zoologically the Mock-turtle'. The Mock-turtle will be classified psychologically, not zoologically, for the scientist in question in these cases is the psychologist, not the astronomer or the zoologist, and if he approaches his task in a 'scientific' way he will concern himself, at the present state of knowledge, not with such creations as these, any more than the physicist will concern himself with the natural fall of a piece of paper, but with far more elementary problems—as, in fact, the experimental psychologist now does—and wait in patience for that remote time when the progress of his science will lead his successors to an understanding of Shakespeare's genius in all its detail.

4

One other point of clarification. On page 236 you say that Sir Cyril Hinshelwood 'gives no support to your argument that the mind-body problem is misconceived, and that, rightly viewed, there is nothing to be discussed'. Evidently I have failed to make my meaning clear, for it is totally foreign to my view that there is 'nothing to be discussed'. I do not for one moment deny that what is often described as the mind-body problem — or, in Hinshelwood's more general phrase, 'the mind-matter relation' —is a real and most important problem: what I think indicates a futile approach to it is that description (except, of course, as a mere label, which I think is the way in which Hinshelwood uses it. It would be pedantic to object to the term 'sunrise', but futile to try to explain the fact on the assumption that the Sun literally rose).

I think there can be no doubt that the solution of the mind-matter problem will come, if at all, through the development of physics and psychology independently until their phenomena can be expressed in common terms—just as the solution of the 'heat-motion' problem came by the development of thermal science and mechanics until they could both be expressed in terms of common concepts, particularly that of energy. We then had a single science of thermodynamics. But if that is true, then we must take account of the fact that 'matter' is not a concept of physics. Physics uses such concepts as mass, time, velocity, electric charge, gravitational potential, and so on, all of which are represented by symbols in its equations and have precise meaning, but 'matter' is not represented by any symbol, and physics would be quite unaltered if the word did not exist. 'Mind', it is true, still appears in much psychological literature, but that is because psychology is a much less developed science than physics, and I think it probable that when it advances further it will find such a comprehensive term as 'mind' to be much too coarse for its purpose, and will discard it. But, be that as it may, it is impossible to express our present knowledge of physics in terms of such a crude concept as 'matter', and if the problem in question is regarded in such a way that that concept must be employed, then we destroy all hope of solving it. The solution, I believe, will come, if at all, when we are able to form concepts general enough to reduce to physical concepts in certain circumstances and to psychological concepts in others, just as the concept of energy is applicable both to the visible motions of large-scale bodies and to the heat which we postulate to account for the quite different sensations we experience when such motions affect our skin through friction. I agree entirely that 'the mind-matter relation cannot be ignored in an intelligent consideration of existence' — I would indeed consider that a masterly understatement—but I do not believe that the solution, when it comes, will appear as a specific relation between 'mind' on the one hand and 'matter' on the other, or that it will ever be reached if we are not prepared to think in terms of scientific concepts rather than the everyday notions that serve us well enough for more superficial ends.

5

And now for the prospects of a harmonization of philosophy, science and religion. I agree with you on the great need for this; I agree in thinking that it is possible; and I share your optimistic view concerning its ultimate realization. But I differ sharply regarding present trends; I think these are ominous in the extreme. My optimism arises not from the signs of the times, but from the fact that the signs of the times are usually misleading, and I have faith, which I can only imperfectly justify, that somehow our present drift will be arrested, and 'the old barbarian' once more frustrated.

> In what wise men shall smite him
> . . . .
> My vision saith not; and I see
> No more.

Between a philosophy whose 'aims and methods' stimulate the problems on pp. 83-4, a science of blind, unbridled mathematical automatism, and the two and seventy jarring religious sects, what chance is there of a reconciliation? None indeed, and in that fact lies perhaps ground for hope, for if these powers of darkness could join forces the situation might well be desperate. But before seeking signs of encouragement, let us look on the black side. I will consider only the immediate outlook for science, of which I know most.

6

Although this is pre-eminently a 'scientific' age, the emphasis at the moment is on the *application* of science, not on the *understanding* of it. We want to know how to do a thing, not what it is that we are doing. This is comprehensible, but it is a highly dangerous situation. We are a people not only driven by forces which we do not understand but also unaware that we are so driven. We respond automatically to the pressure of the moment, ignorant alike of our impulse and our goal.

This is a recent phenomenon. Science in the modern sense of the word dates from the seventeenth century. Its early progress was slow, and it was not difficult for men of the eighteenth century, and most of the nineteenth, to know not only the point they had reached but also how they had reached it. They could see the process as a process, watch its tendencies, and check the stability of its foundations by their knowledge of how those foundations had been laid. For science, unlike other human activities, rests essentially on its past. In music, in art, in philosophy, in religion, in literature, indeed in everything else, it is always possible for some new movement to start whose credentials do not depend on what is past, although its author may be inescapably indebted to his predecessors for its initiation. Beethoven's music would have the same status if Bach were forgotten; Spinoza would not be diminished if we had nothing of Plato; Christianity would remain what it is if all records of Judaism were lost. But it is not so in science. Every scientific paper contains references to earlier papers, each of those to still earlier papers, and so on back to the seventeenth century. Each is valid only if each of the papers to which it refers is valid: let there be an error in one of them, and all that is built on it is vulnerable and in time will collapse.

In our day the structure is so vast that no one who is engaged in pioneering work can give the foundations a thought. He trusts that they are firm, and goes on building. The result is that the bulk of science lurks like a vast 'unconscious', directing to an unknown degree the actions that are consciously performed and, in so far as it contains errors, causing neuroses of which we are now unaware but which are bringing confusion and may bring catastrophe.

This, I repeat, is a modern phenomenon, and its potential danger is modern also—and alarming to those who are aware of it. When science was so elementary that it could be seen as a whole, the danger did not exist. When that first ceased to be possible there was still the ultimate safeguard that experiment would inevitably sooner or later expose an error. That is still true, and many scientists, I fear, rest in the comfortable belief that it is still sufficient. That is a ghastly illusion: an error-exposing experiment now is as likely as not to decimate a nation. We can no longer afford to let our blunders correct our errors: we must discover the errors before the blunders are made.

Considerations such as these have led some of the farther-seeing scientists in several countries to realize the need of studying the whole process of science, with a view to understanding, as far as may be, its essential nature, its possibilities and its hazards. Since the first world war, and more particularly since the second, the subject which has become known as 'the history and philosophy of science' (this cumbrous name indicates the confusion that exists regarding the scope of our intellectual adventures) has advanced its claim to be recognized as a necessary complement to the practice of scientific research. It is essentially a subject in its own right, and yet inseparable from science itself. It is also essentially a single subject. The history of science, regarded as a simple narrative of what has happened without attempt at interpretation, is an interesting pastime, but little more. A philosophy of science based on the present state of scientific knowledge without regard to its ancestry is worse; it inevitably leads to error by petrifying into a logical system what is essentially a fluid process.

As you know, the pioneer in the systematic study of this subject in this country was University College, London, and that is still the scene of the most comprehensive courses. Its achievements, though of course falling far short of the need, have been considerable, and its possibilities are great if a recent tendency, which I note with regret, to separate the 'history' from the 'philosophy' of science, and to emphasize the 'rights' of the former at the expense of the latter, is resisted. The situation is too serious for such petty-minded partisanship. It is regrettable that the subject has not received an indivisible name, but owing to an accident of history—the pioneers in its study, though recognizing its unity, were necessarily unaware of the vital importance which that unity would assume — its two aspects were in most places pursued independently of one another, and their organization on an international scale began independently also. This has now been remedied—at least potentially. The international unions for the history and for the philosophy of science are now linked to form a single International Union for the History and Philosophy of Science, which is one of some dozen or so organisations recognized by the International Council of Scientific Unions (ICSU). The relations between the components are at present rather tenuous, but at least the means exist by which they can be strengthened and improved.

The position of our own country in this matter is embarrassing to those who (by courtesy, not by right) attend the triennial gatherings of the Union and try to explain the non-affiliation of this country in terms which are both truthful and uncondemnatory of one's compatriots. For the fact is that Great Britain is the only major country in the world that is not a member of the Union, and this Union is the only one belonging to ICSU of which it is not a member. The reason is that membership is normally established through the leading scientific organisation of a country. In our case, of course, that is the Royal Society, and the Royal Society is not persuaded that the subject is of sufficient importance to merit such action. Representations have repeatedly been made to it to remove what some of us regard as a stigma on our reputation, but without avail. I have myself, at its invitation, attended a number of meetings of its International Relations Committee at which the question has been discussed, and while it would be improper for me to reveal the reasons given for withholding affiliation, I betray no confidence in saying that they have made me marvel at the strength of the correlation between success in one's own speciality and blindness to everything outside it. There has been no decision not to join. The matter is kept 'under consideration'. It has been so for many years, and there is no prospect that its status will ever be changed.

Not that there is no concern about the matter in this country. The contributions to the understanding of science by individuals here will stand comparison with similar contributions elsewhere, and what such individuals, with their limited means, can do to improve our situation has not been left undone. About fourteen years ago the British Society for the History of Science, with a Philosophy of Science Group as an integral part of it, was formed, and it is still in existence. For purely practical reasons the Philosophy of Science Group was re-named the British Society for the Philosophy of Science, so there are now two societies, independent of one another so far as administration is concerned, but having a large common membership and standing in friendly and mutually helpful relations with one another. Both societies are active, with membership of 180 for History and 270 for Philosophy. They hold regular meetings, and regularly issue publications which command attention in all parts of the world.

These societies—or this Society, as I will call it, since, for most of its history, that has been the correct title—fulfils its function to the best of its ability, and is regarded with respect wherever interest in the subject exists: no adverse criticism of it has ever come to my knowledge. Apart from generous provision of accommodation for its meetings by the Science Museum and University College, London, it has no assistance to supplement the contributions of labour and money made by its members. Nevertheless, its standing is such that when, a few years ago, I was Vice-President of the International Union for the History of Science (a considerable irregularity, since Great Britain was not a member of the International Union, but it indicates the strong desire which the Union had long expressed for this anomaly to be rectified), I was repeatedly told that the British Society for the History of Science would be gladly recognized as representing Great Britain for the purpose of affiliation with the International Union, since the Royal Society was not prepared to do anything in the matter. This, however—quite apart from the extra financial burden of the affiliation fees and from absence of funds to assist delegates to attend meetings in other countries—could not have been carried through without a sense of shame on the part of those who might attend the international gatherings as representatives of Great Britain. Where delegates from the other major countries would be sponsored by their leading scientific organisations, Great Britain alone would be represented by a small new Society, thus inevitably giving the impression that the people of this country are not yet aware of what is happening in the world. Rather than so falsify the real concern felt by large numbers of our countrymen for the study and understanding of the scientific movement (as expressed, for instance, by the resolution of the Convocation of Canterbury which you quoted on page 228), the British Society decided to continue its efforts, as opportunity might arise, to induce a change of attitude on the part of the Royal Society.[1]

To complete the story I was about to add that the British Society had never received support or encouragement of any kind from the Royal Society, either at its formation or subsequently, but that would not be quite correct. I have remembered that at its beginning the British Society asked if it might use the Royal Society's premises for its meetings, and was told that it might. Unfortunately, however, the charge imposed was prohibitive, and accommodation was sought and found elsewhere.

You may well think that there is something seriously wrong

---

[1] Since this was written a further development has taken place. The Royal Society has at last decided to apply for the admission of the United Kingdom to the History of Science Division of the International Union, and the application has been accepted. It has appointed a 'National Committee' 'to maintain contact with the Division of the History of Science of the International Union for the History and Philosophy of Science with the object of facilitating international co-operation in the subject, to recommend to the Royal Society such proposals for scientific action or matters for discussion as may be desirable to bring before the General Assembly of the Union, and also to select delegates to represent the Royal Society at the General Assembly of the Union'. Of the six Fellows whom the Royal Society has nominated to serve on this Committee, four have not previously evinced sufficient concern for the History of Science to join the British Society for this subject (which itself has been invited to nominate one member), so it seems unlikely that the wealth of 'proposals for scientific action' will prove an encumbrance.

In view of the continued indifference of the Royal Society to the Philosophy of Science (to the study of which subject it has made, and intends to make, no contribution of any kind whatever), the British Society for the Philosophy of Science has requested the British Academy (a body bearing a similar relation to the study of non-scientific subjects to that of the Royal Society to science, and, though much younger, of comparable standing in its own sphere) to apply to the Philosophy of Science Division of the International Union for the admission of this country as a member represented by the British Society for the Philosophy of Science. This has been done, and the application has been successful. All financial obligations will fall on the British Society for the Philosophy of Science.

We thus have the farcical situation that one half of a single International Union is recognized by our scientists, and the other half by our non-scientists, neither group otherwise taking any interest in the matter. This is the present extent of the Royal Society's response to the challenge, presented to it by the modern age, to understand what the scientific process means.

here, and may wonder how so unsatisfactory a situation can exist. The reason, I think, is not difficult to perceive, and it lies in a state of affairs which portends the greatest danger for the future. The Royal Society is entirely in the hands of its Fellows. To become a Fellow—a distinction now regarded much more as a reward than as a responsibility—one must achieve something notable in scientific research. It is naturally the aim of the ablest young scientists to gain admission to the Royal Society, but so great is the present competition, and so complex is modern science, that one's chance is slight unless he concentrates exclusively on some narrow field from the very beginning of his career. Anyone who tries to see that field as part of a larger region of interest, even of purely scientific interest, and tries to understand how it affects and is affected by activities which do not form part of it, may as well retire from the competition at once. There are fortunately a few such people, even among the younger scientists, but they will never enter the Royal Society, and so will be without influence in matters on which they are best qualified to make contributions of value. There are also some existing Fellows, elected in more leisured times, who believe that the history and philosophy of science should be a concern of the Royal Society; you have quoted one or two. All honour to them, but they are few, they are among the older Fellows, and there will soon be none left. A study from this point of view of the list of Fellows elected since the last war will cause one with imagination to shudder. A distinguished (foreign) historian of science who had been granted permission to use the Royal Society's library (which contains a wealth of essential literature in the history of science) recently remarked to me that during his whole, rather lengthy, period of study there, not one Fellow of the Society entered the library to consult the older books. If the present attitude to these things is to continue, it would be in the interests of science if the library were offered for sale in the United States. Our displays on ceremonial occasions would be impoverished, but the library would become conveniently accessible to far more of those qualified and eager to use it, and the financial return would be great.

I think it is difficult to exaggerate the seriousness of this position. The scientists into whose hands the control of science in this country is rapidly passing are those who are automatically conditioned to see and understand nothing outside their own speciality. If the only remedy lay in a complete rejection of specialization, it would, I think, be unattainable. Human minds being what they are, specialization at the present stage of knowledge is necessary if progress is to be made. We cannot avoid it, but we can use it intelligently, seeing the parts as parts, but with a feeling of the whole. After all, nature is a specialist. She makes an Aristotle or a Leonardo once in a millennium; all her normal products can do some one thing better than others, and it would be no gain to turn a good mathematician into an indifferent jack-of-all-trades. But we need to recognize that the understanding of science is itself a speciality, and that unless it is committed to the care of those qualified by nature and equipped by training to undertake it, and unless their work is acknowledged to be as important to science as that of the mathematician, we run the hazard of a people who refrain from making traffic laws in the fond belief that that matter can safely be left to the skilled racing motorists who know so much more about the possibilities of engines than anyone else. I see no sign that the Royal Society will come to such a recognition. I think that the hope which Sir Cyril Hinshelwood's wise words have inspired in you comes from your natural assumption that they will have been understood by those to whom they were addressed. I fear that that is not the case. They will have been apprehended as sounds or as marks on paper, but to the great majority they will have conveyed no meaning. They will have been accepted as the proper sounds to make on such an occasion, and worthy to be embalmed and preserved, but that is all.[1]

[1] Lest this should seem a jaundiced view, hear the *ipsissima verba* of a fairly recently elected Fellow of the Royal Society. Professor Fred Hoyle, with that charming naiveté that makes him so lovable, writing in *The Observer* of January 8th, 1961, on 'The essential incoherence of a scientist in embryo', says this : 'For the able [scientific] boy, already certain of his approximate vocation, specialization cannot come too early. Intricate techniques have to be learned so thoroughly that they become second nature—exactly as with a young musical virtuoso. Try to instil an appreciation of the arts into a young scientist, to "round" him, or "broaden" him, or otherwise alter his cultural shape, and your efforts will very properly be wasted. An impassioned lesson on the beauty of the English language will fall on deaf ears—the boy will be pondering some geometrical theorem or perhaps the behaviour of a new brand of bug.' Hoyle goes on to argue that such boys, after maturing, should have a greater influence in the government of the country because of the valuable ideas which they alone can contribute. For example : 'To the scientist, war starts because human behaviour is representable in terms of mathematical equations possessing discontinuous solutions'.

I am sure Hoyle is right in maintaining that no one but the 'scientist' would

7

I will not attempt a similar estimate of the outlook in philosophy and religion as I have no comparable knowledge of those fields, but even a superficial examination is sufficient to show that the prospect here is little, if any, brighter. In one important respect it is perhaps worse. There is now a fairly general consensus of understanding — among those who value understanding — of what we mean by science, though we are still some way from unanimity, but with regard to philosophy and religion the most diverse views are held. What is philosophy? What is religion? Many answers can be given to these questions, all having weighty support and all incompatible with one another. One cannot begin to consider the possibility of a synthesis of philosophy, science and religion until one knows what the words mean, and since there are no natural meanings of words, but only conventional ones, there is no authority to which appeal can be made. Rather than make an arbitrary choice, I think it is better temporarily to discard all such terms, and attempt a survey of the whole field of human endeavour to see if a classification of its elements can be made in which there is no overlapping and no inevitability of discord. It might then be possible to attach the names, *philosophy, religion, science,* to three such classes, but that will be relatively unimportant. The fundamental question is whether a complete and harmonized life is possible, for the individual and the whole community of individuals.

Thinking along these lines, I would suggest a fundamental division between what I shall call *experience* and *voluntary action*. By experience I mean that of which we are aware, that which is given to us, so to speak, without our having designed it and independently of any wish of our own. It includes not only sensations—sights, sounds, smells, etc.—but all feelings of pleasure and pain, emotions and passions and impulses and so on. Voluntary action, on the other hand, is what we choose to do and could avoid doing if we would. Of course, the two things are often associated with one another. I might choose to look at the

---

have thought of this. But he is right also in holding that this is the type of young scientist who alone can now expect to reach a position of influence. Fortunately, it is not too late to keep him out of the Government, but unless something drastic is done it is virtually certain that in a few years' time the Royal Society—and therefore the chief control of science in this country—will be entirely in his hands.

sky to see the stars, but my choice here is merely that of opening my eyes and turning them in a certain direction: what I then experience is not of my contrivance. From previous experience I might have a confidence, almost amounting to certitude, that I shall see a particular pattern of stars, and I may, in fact, choose to look into the sky because I want to see that pattern, but nevertheless, what I see is given me and is independent of my desire. The stars may be obscured by a cloud, or the pattern may be spoilt by the advent of a new star. All that is independent of my will; it is part of my experience, and whether I see points of light or uniform greyness, and feel pleasure or disappointment, I have to accept it as something outside my control.

I have spoken before (p. 196) of three possible attitudes to experience. The only one that concerns us now is that of what I there called 'the scientist and the philosopher': the names are now in suspense, but the attitude remains, namely, that of using our reason to construct a rationally connected system of concepts which stand in a relation of one-to-one correspondence with experience — in brief, the rational correlation of experience. Reason is a third element of our being, neither a matter of voluntary choice (I do not choose that 7 is a prime number and 8 is not; I simply have to accept it) nor of experience—though, like experience, it is 'given'. In a full analysis, of course, the distinction between experience and reason would have to be made precise, but this is a discussion, not a complete treatise on all that can be talked about, and I think the distinction will be understood readily enough. As a rough working rule, experience is temporal — it can always be located in time and may cease. Reason, on the other hand, is timeless. A prediction that the Sun will rise on this date next year may conceivably be falsified—that is a matter of experience. A prediction that, this being the year 1961, next year will be 1962, cannot be falsified; we do not have to wait until next year to see if it comes true. That is a matter of reason.

The whole field of voluntary choice lies outside this. Experience and reason are given to us and have to be accepted: which of a number of alternative *actions* I choose on any particular occasion rests with me. This is quite independent of any theories about freewill or determinism. I have the present consciousness that I can go on writing or stop and listen to the wireless programme. Whichever I do, it is possible to say that it was determined beforehand or it was not; that does not alter the fact of my now having to make a decision as to what to do. I do not choose whether twice two shall be four or five: that is a matter of reason which I must accept. I do not choose whether I shall see or not when I open my eyes: that is a matter of experience, which also I must accept. But I do have a choice, not present in these cases, whether I write or listen, and, no matter whether my choice is 'determined' or not, it does exist in this case and not in the others. Such situations, of which this is a merely trivial example, exist throughout life and compose a whole field of action which lies altogether outside reason and experience according to the definitions I have adumbrated.

We can now venture to assign names. Let us call the rational correlation of experience, *science*, and the commitment to a particular course of action, *religion*. *Philosophy* can be what I am trying to practise now, namely, the attainment of a point of view from which the whole of consciousness can be classified in such a way that no contradictions arise, and the various problems and loose ends which of necessity such a cursory survey leaves undisclosed become susceptible of definitive resolution.

To save misunderstanding I should say that this nomenclature does not wholly accord with that used in my book, *Through Science to Philosophy*, to which you kindly referred earlier. There I regarded science as that which is now usually so designated—astronomy, physics, botany, etc.—and philosophy as the contemplated extension of science over the whole field of experience. In terms of my present dictionary, the earlier 'science' is the present state of science, and the earlier 'philosophy' is completed science. Also, the definitions I am now proposing are often at variance with the senses in which I have used the words in question at earlier stages of our discussion, where they have necessarily had to bear their customary meanings—or, rather, each has had to bear that one of its customary meanings which the context has demanded. But where so much confusion prevails, consistency is incompatible with intelligibility. I trust that it will be understood that I am not now saying that philosophy, religion and science *are* such and such—a meaningless statement, since we choose words to represent things, they are not things given us to elucidate—but merely using those words to designate a particular classification of the whole field which, in other uses of them, is classified differently.

All that has gone before makes it unnecessary for me to say anything more about science. My definition here accords more closely with that usually implied by the word than it does in the other cases, and the practitioners of science are, I think, on the whole more faithful to their trust than are philosophers and followers of religion—perhaps because of the more clearly defined nature of their task. When they fail it is more often in bringing the wrong offering to their true god than in turning to false ones. They tend to construct logical systems — though with insufficient grounding in logic outside the routine processes of mathematics—for their own sake, and force them into an impossible correspondence with experience by invoking hypothetical experiences that have not occurred; but they do not, like some philosophers of old, claim that, when the correspondence is found to break down, the experience must be denied. Nor need I say anything more about philosophy, since it is what, well or ill, I am engaged in now. What nowadays usually goes by this name is either linguistics (or semantics or whatever title pleases its ear)—on that I have said enough—or 'existentialism', about which I can say nothing since I do not know what it means. As a practical rule, I have found it very useful, whenever I come across the adjective, 'existential', to cross it out, when the sentence usually takes on a meaning, but I have not succeeded in finding a similar device for 'existentialism'. But religion calls for more comment.

8

Religion I have defined as concerned with the field of voluntary action. It therefore includes—or dispenses with, as one pleases—ethics, since that also is an attempt to deal with this field. We are all, of course, choosing among a number of alternative actions at every moment of our waking lives, so if religion is merely choice, everyone is religious. But there is a distinction between choosing on each occasion 'on the spur of the moment', without regard to the choice on any other occasion (no matter how 'consistent' a person's actions may turn out to be when studied by a developed science of psychology), and the deliberate devotion of one's whole life to a particular single end; and I would reserve the word 'religion' for the latter. Religion is then obedience to the great commandment expressed by Judaism and Christianity as, 'Thou shalt love the Lord thy God with all thy heart, and with all thy soul, and with all thy mind, and with all thy strength'. In other words, religion is unreserved dedication, commitment, devotion of one's whole being to a single ultimate purpose. This act of dedication, performed without ceasing, is worship. The difference between one religion and another lies in the choice of object of worship, in the content of the phrase, 'the Lord thy God'.

As thus defined, the distinction between the so-called 'religions'—Buddhism, Hinduism, Christianity, Judaism, Islam, etc.—may be secondary: apart from matters of ritual, they often differ chiefly in 'creed' or 'theology', which is a matter of science, good or bad; there is less difference in the character of their God. Hitler was an intensely religious man, but his God was not ours. Absolute commitment is, of course, practically, if not theoretically, unattainable; we simply do the best we can, and this is a case where the attempt and not the deed does not confound us. The God of Christianity is defined, so far as definition is possible, by the second commandment: 'Thou shalt love thy neighbour as thyself'. Love here, of course, does not necessarily imply any particular emotion, the feeling one has for one's sweetheart or family, for instance, though something like that may sometimes be associated with it. I am not conscious of any feeling of sentimentality towards myself, but I automatically tend to seek what I believe to be my own welfare: I do not doubt that others are much the same as I in this respect. To love them all as I love myself imposes such a restriction on my instinctive impulses as to determine the whole course of my life—or would do if my commitment were absolute. There is no room for any subsidiary rules or articles of belief that could conceivably conflict with anything in science; all the law and the prophets are here included.

The word 'religion' nowadays almost invariably includes what we call ethics and theology. I would exclude them both—the first since it is at best superfluous and at worst a hindrance, the second as coming within the field of science. Absolute commitment to a single end necessarily excludes the prohibition of any means by which that end can best be achieved. To the really religious man all things are lawful, as to St Paul and St Augustine. Of course, the mere fact that a particular kind of conduct is normal and is embodied in the laws of the State is a tremendous factor in determining what one's religion requires him to do in any particular situation, and it is almost invariably a decisive factor, so far as it goes, but it is not paramount. In the last resort it is to be followed only when the consequences of doing otherwise, so far as one can estimate them, are on the whole worse—which is usually the case. Also, on a more superficial level, rules of conduct may be followed for practical reasons. Anyone may know of failings to which he is prone, and adopt a particular regimen for their correction, but that is a very different thing from the universal 'Thou shalt not' of most ethical systems. Many very worthy people regard such principles as vegetarianism, anti-vivisectionism, pacifism, teetotalism, and so on, as essential elements of religion, but they are precisely the opposite: they are automatically excluded from religion as defined by the first commandment. Since they are all negative they cannot themselves become objects of worship—you cannot dedicate your whole being merely to not doing something—and in so far as they are coupled to a positive dedication their only possible effect can be to qualify it and so to destroy the absoluteness which is its essential character. You serve God unreservedly—unless His service requires so-and-so. No one knows what situations may arise in life, but—taking Christianity as an example—we have overwhelming evidence that it is frequently impossible to promote the welfare of some without retarding or destroying that of others, and in such a situation the prohibited act becomes perhaps inevitable and in any case demanded by the Christian religion.

Theology is an attempt to form a concept of God which serves to make sense of experience in exactly the same way as does the concept of the atom or of the unconscious mind. It is therefore of essentially the same character as science, but the experiences concerned are chiefly those which have not yet come within the scope of the generally recognized sciences. Sometimes a theology unwisely attempts to cover experiences already better dealt with in an existing science — as in the account in *Genesis* of the formation of the physical world, for instance—and it is then usually discredited sooner or later; it is simply bad science. More often nowadays a theology is an attempt on the part of an individual to synthesize his own personal experiences—sensations, which are paralleled by experiences of most human beings, and mystical experiences, which come to fewer and in detail perhaps to no others—into a single system. As a personal adventure this is entirely unobjectionable: as an objectively valid scheme of thought it is most unlikely to achieve permanence.

If one wishes to conceive the object of his worship in the form of a personal Being—i.e. of a man writ large, since that is the only way in which we can possibly conceive of such a Being—then the only human attribute that one can properly ascribe to God is that of being able to commune with and sustain the individual in the working out of his commitment. Such a conception, based firmly on experience, is invulnerable to science or anything else, and can only be included in, and not destroyed by, any more complete conception that may later become possible. And indeed, in their best moments religious thinkers have emphasized the utter inexpressibility of the being of God. But unfortunately it is far commoner for a theologian to credit his God with all the attributes we at present believe to belong to ourselves. We talk of the Will of God—meaning the choice which he makes between alternative ways of running the world; and, further, we often know what God's will is, and give reasons for holding it to be such as to merit our commendation.

There was quite an outcry a short while ago when the Archbishop of Canterbury, if I remember rightly, happened to say that it might be the will of God that the human race should perish and make room for some higher form of being. People who were sure that they knew the will of God better than that even lampooned him in cartoons. We can imagine a similar protest among the dinosaurs if one of them had happened to opine that perhaps the mammals might be intended by the Supreme Designer to supersede their kind. When our religious people so far transcend the limits of experience as this, a conflict with science, in which they must lose, is inevitable, because they are simply becoming bad scientists. Experience must be accepted, whether we like it or not, and if we wish to form a conception of its cause—no matter whether we call that cause God or nature or necessity or anything else — it must stand the test of experience in exactly the same way as any other scientific hypothesis.

The religious body to which I belong—the Society of Friends—has neither ethical system nor theology, although any individual member may have his own system of belief. One's guide in conduct is the Inner Light, which is not expressible in written rules, and meetings for worship are in silence since in the last resort there is nothing that can be said. This seems to me to be the essence of religion. Of course, this Society, like any other human institution, falls short of its ideal. It traffics with pacifism, and—in that connection particularly—makes categorical statements about God's intentions which suggest the intrusion of some unacknowledged Outer Light. But it comes nearer than anything I know to the heart of things, to the point of view from which the whole of life can be seen to form a possibly harmonious system in which one's intellect and one's actions are both completely free to be dedicated to the ultimate ideal without fear of conflict. And the characters it produces—the best of them—seem to me to come nearer to the fulfilled life than those found in the same proportion in any other organized body.

9

In the foregoing paragraphs I have outlined the field of our discussion in terms which seem to me to offer a possibility of a satisfactory synthesis. But possibility is one thing and probability another, and I have no expectation that it will lead to a sudden amelioration of our present discontents. The divisions of thought which it requires, the terminology which it employs, and the point of view from which the survey is made are all too unfamiliar for that, and familiarity in these matters is another name for inertia.

We need not look far to see how tenaciously men will cling to accustomed classifications, even in the most desperate situations. How often, when the trend of science seems to support a particular theological belief, have we not heard science acclaimed as the arbiter whose decision cannot be disputed! When, as invariably happens, the next scientific advance shows the matter in quite a different light, the attitude changes: science is then reproached or ignored. Appeal has been made to Caesar and his ruling accepted when it is favourable, but when it is not his authority is denied: this is dishonest. Yet it is unavoidable if beliefs concerning the origin of matter are held to be a part of religion. On the other side things are no better. A modern cosmologist has felt compelled to prefer his view of creation to that of the writer of the Book of Genesis, and thereupon concludes: 'It seems to me that religion is but a desperate attempt to find an escape from the truly dreadful situation in which we find ourselves'.[1]

[1] F. Hoyle, *The Nature of the Universe*, p. 100 (Blackwell, 1960).

Notwithstanding all this, many whose religion is the most precious thing in their lives find it almost impossible to divide what they have regarded as a single thing into parts. It appears to them sacrilege to describe some of their cherished convictions as speculative cosmology and others as rudimentary psychology. They are shocked when they hear what they call the working of God in the soul described in terms of 'the unconscious mind', even though what is so described remains exactly the same. *Pace* Juliet, names are very powerful things.

Do you know the story of the three men facing a loaf of bread? 'We call it "pain"', said the Frenchman. 'We call it "Brot"', said the German. 'We call it "bread"', said the Englishman, 'and it *is* bread.' We all laugh at this, but consider the phrases, 'historical necessity' and 'the invincible purpose of God revealed in history', as referring to the course of past events. Whether or not what they designate is a reality or an illusion— that is quite irrelevant for our present purpose — the phenomenon in question, and the evidence for it, are precisely the same whichever phrase is chosen. Yet I do not think that either the average theologian or the average materialistic historian will see the choice merely as a matter of convenience.

I therefore cannot underestimate the difficulties facing the acceptance of such a classification as I have suggested, and I am fully aware of the prime source of those difficulties, which lies in the fact that that classification implies that we start from consciousness rather than from an external universe presented to consciousness. (I use the word *consciousness* here to cover a wider field than experience, which was sufficient when we were discussing scientific problems. In consciousness I include experience, reasoning, volition and anything else that makes up our total 'aliveness'. I do not here use the word in the limited sense in which, in psychology, it is the correlative of 'the unconscious'. I am sorry for this necessity of narrowing or extending the meanings of familiar words, but it *is* a necessity since their customary significance is indefinite.) You start with such a universe, and therefore with the great advantage that you can speak the language of ordinary intercourse much more freely than I can. You can present the problems in a more familiar way, and so appear to be more directly in contact with them.

My approach by comparison seems academic, and such conclusions as it may lead to seem unreal: you appear to be building on a rock and I on sand. All I can plead in reply is, first, that the sand is indubitably there while the rock has to be postulated or precariously inferred; and secondly, that although the problem originally presents itself to us in your terms, and the solution must also probably be so expressed in order to be acceptable, the passage from problem to solution is possible in my terms but not in yours. (Incidentally, when I was in Israel a few years ago, I was told that buildings of fifteen storeys could now be erected with perfect safety on the sand which covers much of the land there, so I take heart.)

I can illustrate what I mean by a parable of Eddington's, which I think is more illuminating than his two tables:

'The learned physicist and the man in the street were standing together on the threshold about to enter a room.

'The man in the street moved forward without trouble, planted his foot on a solid unyielding plank at rest before him, and entered.

'The physicist was faced with an intricate problem. To make any movement he must shove against the atmosphere, which presses with a force of fourteen pounds on every square inch of his body. He must land on a plank travelling at twenty miles a second round the sun—a fraction of a second earlier or later the plank would be miles away from the chosen spot. He must do this whilst hanging from a round planet head outward into space, and with a wind of ether blowing at no one knows how many miles a second through every interstice of his body. He reflects too that the plank is not what it appears to be—a continuous support for his weight. The plank is mostly emptiness; very sparsely scattered in that emptiness are myriads of electric charges dashing about at great speeds but occupying at any moment less than a billionth part of the volume which the plank seems to fill continuously. It is like stepping on a swarm of flies. Will he not slip through? No, if he makes the venture, he falls for an instant till an electron hits him and gives a boost up again; he falls again, and is knocked upwards by another electron; and so on. The net result is that he neither slips through the swarm nor is bombarded up to the ceiling, but is kept about steady in this shuttlecock fashion. Or rather, it is not certain but highly probable that he remains steady; and if, unfortunately, he should sink through the floor or hit the ceiling, the occurrence would not be a violation of the laws of nature but a rare coincidence.

'By careful calculation of these and other conditions the physicist may reach a solution of the problem of entering a room; and, if he is fortunate enough to avoid mathematical blunders, he will prove satisfactorily that the feat can be accomplished in the manner already adopted by his ignorant companion.'[1]

Who would be a learned physicist? But suppose the man in the street is approaching the door with a suitcase in each hand which for some reason he does not wish to release. It is he who is now faced with an intricate problem. But, thanks to the learned physicist, doors can now be made (they are commoner in the United States than here) which automatically open when one approaches them and close again after one has passed through. How has this become possible? Not by calling a door a door, though both the problem and its solution are completely expressible in those terms, but by forgetting all about doors and studying, quite independently of one another, the phenomena of electricity, magnetism, optics, mechanics, and so on, and ultimately bringing them together into a synthesis which enables us, among hosts of other things, to make self-opening doors.

I do not believe that the problem of harmonizing what we now call philosophy, religion and science will be achieved without ignoring the indefinite entities now bearing those names and making a fresh analysis of consciousness such as I have here attempted, or some better one, which will be free from confusion and overlapping.

Nor, I think, will it be necessary to wait for a general realisation of this before it is put into practice. The process by which the self-opening door was made possible presupposed a division of experience very different from that ordinarily made, but this has only recently begun to be realized. Throughout the history of science from the seventeenth century onwards we have been studying the laws of motion, heat and the rest quite independently of one another, and all the time thinking we have been studying the bodies believed to possess those properties. I have described this process elsewhere,[2] and need not repeat it here. Its

---

[1] *Science, Religion and Reality*. Edited by J. Needham (Sheldon Press, 1926), p. 189.
[2] *The Scientific Outlook in 1851 and in 1951*. [Brit. Journ. Phil. Sci., Vol. II, p. 85 (1951)].

relevance for our present purpose is that it shows that we need not know exactly what we are doing in order to do the right thing, and therefore we may not necessarily have to await the acknowledgment of such an analysis as I have suggested in order to proceed on the basis of it. The difficulties may thus not be so great as they appear — though this, of course, in no wise diminishes the dangers of failing to study what it is that we are doing.

That this may not be entirely a false hope may perhaps be seen by considering just one example. One of the best loved of our hymns, by Isaac Watts, contains the verse:

> Were the whole realm of nature mine,
> That were an offering far too small;
> Love so amazing, so divine,
> Demands my soul, my life, my all.

That is a strange thing to say from the 'realistic' point of view, which Watts almost certainly took for granted. What right has he to imply that his soul is far greater than the whole realm of nature? Where was that soul when the foundations of the earth were laid? Had it commanded the morning, and entered into the springs of the sea? I have no doubt whatever that Watts would have admitted that he did not then exist, either in the universe as it is in itself or in the universe as it was to him, yet he wrote, and rightly wrote, as though the whole realm of nature was a concept formed to interpret sensory experiences which were only a small part of his total consciousness.

10

Perhaps the fact of this 'blind understanding' contributes to the conviction which, apparently against all the evidence, I feel that we shall somehow emerge from the present confusion into a state of greater enlightenment; though how, in detail, it is to come about, I cannot imagine. I agree with what you said at the beginning, that philosophy occupies the key position—or rather, that the key position is at the moment vacant, and philosophy, if it existed, would be the rightful heir to the throne. Philosophy, as I have defined it—the attempt to describe the whole field of possible thought in terms which allow of a synthesis free from overlapping or disharmony—would appear to be presented with the initial task: when that is performed the actual synthesis can begin. But philosophy in this sense seems to be a forgotten enterprise, and until it is recollected our understanding may have to remain blind. But, even so, we may nevertheless go in the right direction.

I cannot agree with the view, often expressed nowadays, that our advance in science has too far outstripped our moral progress: it has even been proposed that science should call a halt and allow morals to catch up. As soon as I try to give precision to such a conception it disperses into a mist. I cannot conceive of any possible common yardstick by which to measure the two things, and I can give plausibility to the idea only by supposing that those who hold it are unconsciously estimating the position of science by its distance from its starting-point, and that of morals by the distance from the goal. It is then true that we have learnt a tremendous lot, and fall far short of what we ought to be. But to compare our impression of the achievements of one thing with our impression of the shortcomings of another is to do something quite without significance.

I am inclined to think that my optimism arises chiefly from the fact that men seem to be essentially religious—in my sense of the word. That is to say, there is something in men that leads them to dedicate themselves to a purpose, and to pursue it at the expense of narrowly selfish ends, even though they may be unaware that they are doing so or of what it is that they are pursuing.

Take any movement which we regard as a desirable reform—the abolition of slavery, the acknowledgment of the earth's mobility, anything at all that passed from extreme unpopularity into general respectability. How did that happen? There were pioneers, of course—Wilberforce, Clarkson, Copernicus, Galileo—who had to face opprobrium or worse. Their part was indispensable, and they can hardly be overpraised, but all their efforts would have been futile had there not existed in men generally an impulse to respond, even at the cost of such things as economic gain and the intellectual satisfaction afforded by the most complete synthesis of knowledge the world has known. Had the pioneers not acted as they did, in all probability others would have done so, but if there had been no general impulse to respond there would have been no reform. Quakers express this in their belief in 'that of God in every one'—a quaint historical phrase which for many may hide rather than reveal the fact that it denotes a simple truth about human beings which history continually exemplifies.

I am inclined to think, therefore, that in so far as there is any sense in comparing our scientific with our moral state, it is the former that needs catching up with the latter. We needs must love the highest when we see it: the difficulty is to see it. When our knowledge of psychology has caught up with our physical knowledge we shall be in a better state to see what possibilities the future holds and better equipped to realize those that we desire. We shall still, of course, presumably be free to choose our God, our object of worship, our ultimate aspiration. In the meantime we can strive for greater integrity in our thinking, seeking out and eliminating the inconsistencies that lie unrecognized in our ideas.

It is probably my belief in the essential 'healthiness' of human nature that is the basis of my faith that somehow all will be well. Although I cannot see how, I believe that there will again be philosophers, that the Royal Society will recover from the blindness of its present functionaries, and that organized religion will shed its bad science and superstitious taboos; and that all together

> Shall make one music, as before,
> But vaster.

APPENDIX I

1

Note to p. 54, on the question of an Ether

SIR ARTHUR EDDINGTON (in a lecture at Cornell University in 1934) said:

'As far as and beyond the remotest stars the world is filled with aether. It permeates the interstices of the atoms. Aether is everywhere . . . There is no space without aether, and no aether which does not occupy space. Some distinguished physicists maintain that modern theories no longer require an aether—that the aether has been abolished. I think all they mean is that, since we never have to do with space and aether separately, we can make one word serve for both; and the word they prefer is "space". I suppose they consider that the word aether is still liable to convey the idea of something material. But equally the word space is liable to convey the idea of complete negation . . . The essential truth remains. You cannot have space without things or things without space; and the adoption of thingless space (vacuum) as a standard in most of our current physical thought is a definite hindrance to the progress of physics.'

SIR OLIVER LODGE (in a book published in 1925, with the title *Ether and Reality*):

'All pieces of matter and all particles are connected together by the ether and by nothing else. In it they move freely and of it they may be composed. We must study the kind of connection between matter and ether. The particles embedded in the ether are not independent of it, they are closely connected with it, it is probable that they are formed out of it: they are not like grains of sand suspended in water, they seem more like minute crystals formed in a mother liquor. The mode of connection between the particles and the ether is not known; it is earnestly being sought: but the fact that there is a connection has been known a long time. We know it, because a particle cannot quiver or move without disturbing the medium in which it is. A boat cannot oscillate on the surface of water without sending out waves of sound; a particle cannot vibrate in ether without sending out waves akin to those of light . . . Ether is the universal connecting link; the transmitter of every kind of force. Action at a distance is wholly dependent on the ether, and it is manifestly the vehicle or substratum underlying electricity and magnetism and light and gravitation and cohesion . . . The ether is a physical thing . . . Its mechanism is unknown to us, its inner nature eludes us; yet mechanism it must have, for it is subject to physical laws.'

SIR J. J. THOMSON (quoted by Lodge in the same book):
'In fact, all mass is mass of the ether; all momentum, momentum of the ether; and all kinetic energy, energy of the ether. This view, it should be said, requires the density of the ether to be immensely greater than that of any known substance.'

Thomson restated his conviction in a letter, written to a correspondent, towards the end of his life, which is quoted by his biographer, Lord Rayleigh. He wrote: 'I differ from you about the value of the conception of an ether, the more I think about it the more I value it. I regard the ether as the working system of the universe. I think all mass, momentum and energy are seated there and that its mass, momentum and energy are constant, so that Newtonian mechanics apply.'

ALBERT EINSTEIN, in a paper published in 1935 in *Mein Weltbild*, and republished, also in 1935, in an English translation of essays and addresses under the title *The World as I See It*, p. 204:
'We may sum up as follows: According to the general theory of relativity space is endowed with physical qualities; in this sense, therefore, an ether exists. Space without an ether is inconceivable. For in such a space there would not only be no propagation of light, but no possibility of the existence of scales and clocks, and therefore no spatio-temporal distances in the physical sense. But this ether must not be thought of as endowed with the properties characteristic of ponderable media, as composed of particles the motion of which can be followed: nor may the concept of motion be applied to it.'

More recently, PROFESSOR DIRAC of Cambridge, eminent among the mathematical physicists of our time, created something of a sensation by publishing, in *Nature* (November 24th, 1951), a short paper with the title 'Is there an Aether?':

'In the last century, the idea of a universal and all-pervading aether was popular as a foundation on which to build the theory of electro-magnetic phenomena. The situation was profoundly influenced in 1905 by Einstein's discovery of the principle of relativity, leading to the requirement of a four-dimensional formulation of all natural laws. It was soon found that the existence of an aether could not be fitted in with relativity, and since relativity was well established, the aether was abandoned. Physical knowledge has advanced very much since 1905, notably by the arrival of quantum mechanics, and the situation has again changed. If one re-examines the question in the light of present-day knowledge, one finds that the aether is no longer ruled out by relativity, and good reasons can now be advanced for postulating an aether ... Thus with the new theory of electrodynamics we are rather forced to have an aether.'

LORD CHERWELL (Professor F. A. Lindemann), in a lecture in May 1955 said:

'In the old days one talked of waves in the ether, but it turned out that the ether had to have so many weird and wonderful properties that people no longer talked about it and indeed I am tempted to say seem to want to hush it up.' And I remember Lord Cherwell saying to me once in conversation that he could not see that any advantage was gained by using the word 'field' instead of the word 'ether'.

And SIR EDMUND WHITTAKER, the author of the monumental *History of the Theories of Aether and Electricity*, wrote as the concluding words of the Preface to the enlarged and revised edition in 1953:

'As everyone knows, the aether played a great part in the physics of the nineteenth century; but in the first decade of the twentieth, chiefly as a result of the failure of attempts to observe the earth's motion relative to the aether, and the acceptance of the principle that such attempts must always fail, the word "aether" fell out of favour, and it became customary to refer to the interplanetary spaces as "vacuous"; the vacuum being conceived as mere emptiness, having no properties except that of propagating electro-magnetic waves. But with the development of quantum electrodynamics, the vacuum has come to be regarded as the seat of the "zero-point" oscillations of the electro-magnetic field, of the "zero-point" fluctuations of electric charge and current, and of a "polarization" corresponding to a dielectric constant different from unity. It seems absurd to retain the name 'vacuum' for an entity so rich in physical properties, and the historical word "aether" may fitly be retained.'

2

I would add to this Note an example of the inconvenience caused by the veto of the scientific authorities on any kind of recognition of an ether hypothesis.

A new vocabulary has had to be devised for broadcasting, and in it we have been reduced to the absurdity of speaking of broadcast programmes being 'on the air', although everyone knows that the air has hardly anything to do with it. There is at the beginning a transmission of sound-waves in the atmosphere, originated by a human voice, or by musical instruments or whatever it may be, and travelling for a few feet, or perhaps only a few inches, to be picked up by a microphone: and again, at the end another set of sound-waves in the atmosphere between an amplifier in a receiving-set and a human ear, or some kind of recording apparatus. But apart from that the air does not come into the matter at all; the transmission, perhaps for a thousand miles, is not atmospheric.

Lately it has been found possible to beam a transmission from England to the Moon, to be reflected from its surface to Washington, and be heard there by the President. The electro-magnetic pulses have been travelling for almost the whole of their journey far outside the limits of the Earth's atmosphere; yet the engineers who have been doing this are expected to use the absurd phrase of having been 'on the air'.

The point came up for consideration when an official Committee, under the Chairmanship of Lord Beveridge, was appointed, in 1949, to consider the terms for a renewal of the Charter of the British Broadcasting Corporation, soon due to expire. In the Introductory Chapter to their voluminous Report, on page 2, the Committee wrote:

'We have used a few words and expressions in our Report which, although technical in character, are in everyday use and are generally understood. We have spoken, for instance, of the "ether" and of setting up "waves" in the ether, though, as appears from the definition of technical terms given to us by the BBC and printed in part in Appendix F, the ether is no more than a "hypothetical non-material medium, filling all space, the existence of which is postulated for theoretical purposes in relation to the propagation of electro-magnetic waves". The ether, so defined, is a hypothesis rather than a reality, and it may well be argued that there cannot be waves in a hypothesis. We have been informed by high authority that scientists are now able to explain the phenomena of nature to themselves without postulating ether or waves in the ether. But no one appears yet to have found a way of explaining broadcasting to the layman without using such terms. We have followed the established practice. We have been verbal purists only to the extent of not speaking of broadcasts as being "on the air".'

APPENDIX I

Nevertheless, this almost meaningless expression continues to be used by the BBC in announcements and publications. I am informed that the BBC does not hold that it is its function to lay down a rule in a matter on which expert opinion has given no guidance, and it prefers not to establish any convention or issue any instruction to bind its speakers: if there should come about a consensus of scientific opinion that would give a guiding line, the situation might be altered.

There is certainly force in this objection. But meantime the needs of daily usage are evidently having an effect. In May 1959 I received from the BBC a copy of an official Memorandum on a matter of public interest—presumably sent to me as having been for many years a member of its General Advisory Council; and I was encouraged to find that the first words of the first paragraph were—'Broadcast programmes are carried through the ether from the transmitter at the transmitting station to the receiver in the home on electro-magnetic "waves"—through the ether'!

APPENDIX II

Proof that Einstein's Special Theory cannot correspond with fact (p. 73)

Consider a group of bodies all relatively at rest. Choose one of them as the origin of co-ordinates, and place a standard clock there. Then every one of the bodies is at rest in a co-ordinate system thus chosen, and the time of any event that occurs among them is obtained by letting a beam of light proceed from the event to the clock at the origin and subtracting from the time which the clock records at its arrival the quantity r/c, where r is the distance of the event from the origin and c is the velocity of light. This is the definition adopted by Einstein, and there is no reason to quarrel with it.

It is often expressed in a slightly different form. We suppose a standard clock, at rest in the co-ordinate system, placed at each point of space and synchronized with the clock at the origin by the dispatch of a light signal from the origin to this clock and its immediate return. Then the clock in question is set so that it records the time of receipt of this signal as the mean of the times, by the clock at the origin, of its dispatch and return. The time of any event is then given directly by the reading of the clock at the place and time at which the event occurs.

It is obvious that this gives the same value for the time of the event as the former definition, and indeed the same value is given by any clock of this set, if we make the proper allowance for the time of travel of light from the event to that clock. In other words, the time of an event in any co-ordinate system is a property of the *system*, not of any single clock in it. Though this will hardly be questioned at this stage it is necessary to insist on it because many errors have been made in applying the Lorentz transformation (which so far we have not introduced) to actual events, which can be traced to a tacit assumption that the time of an event in a single co-ordinate system varies with the clock, stationary in that system, by which it is determined.

Einstein was perfectly clear on this point. In his original paper on the subject[1] he wrote: 'It is essential to have time defined by

---

[1] *Ann. d. Phys.*, 17, 891 (1905); English translation in *The Principle of Relativity* (Methuen, 1923), p. 40.

APPENDIX II

means of stationary clocks in the stationary system, and the time now defined being appropriate to the stationary system we call it "the time of the stationary system"'. It is obvious from this that the time, in any single co-ordinate system, of any event has a single value, no matter by what clock of the system it is evaluated. Again, Einstein and Infeld, after giving the definition of time in a co-ordinate system in terms of a multitude of clocks as described above, write[1]: 'When discussing measurements in classical mechanics, we used one clock for all co-ordinate systems. Here we have many clocks in each co-ordinate system. This difference is unimportant. One clock was sufficient, but nobody could object to the use of many, so long as they behave as decent synchronized clocks should.'

It is clear, then, that one clock is sufficient, and we shall suppose that that is all that we have—a single clock at the origin that gives the standard time for the system of every event that occurs.

Next, it must be noticed that we say 'the time for the system' and 'every event that occurs', not 'every event that occurs in the system'. An event — a point-instant — is independent of all co-ordinate systems; it is meaningless to speak of it as 'stationary' or 'moving' in any such system. If the event is the instantaneous collision of a large number of bodies, all coming from different directions, it can be regarded as happening to any one of them: they are all moving differently, but the time of the event in any co-ordinate system is exactly the same, no matter on which of the bodies we regard it as occurring. Events are essentially independent of co-ordinate systems, but their places and times may vary with the system chosen.

Suppose now that we have two groups of bodies, all the bodies in each group being relatively stationary and moving with uniform velocity with respect to all the bodies in the other group. We may think of two swarms of stars passing through each other, like the two star streams of Kapteyn, the peculiar motions of the stars in each stream being neglected. Suppose that the stars at the origins pass one another at an instant at which the standard clocks placed on them both read zero, and that at some later time two stars, one in each group, collide. That is an event, and it has a definite time in each co-ordinate system. It is the same event, no matter whether the observer at the origin of stream one regards it as a disturbance of one of his stationary stars or whether the observer at the origin of stream two regards it as a disturbance of one of *his* stationary stars, but the times of the event which the observers record may of course not be the same.

[1] *The Evolution of Physics* (Cambridge, 1938), p. 191.

Let us suppose they are not. Then since the clocks agreed when they were together, they must have run at different rates afterwards, for if we want to compare the rates of clocks the only conceivable way of doing so is to compare the time intervals which they record between the same events. The events belong no more to one co-ordinate system than to the other, but the times that elapse between them are properties of the systems, i.e. of the clocks that record the times of the systems. Hence, if one clock shows, say, twice the time interval shown by the other between the same two events, that clock must go at twice the rate of the other.

Now the clocks are supposed to be of identical construction, so that if one is regarded as running uniformly in its system, the other also must be regarded as running uniformly in its system. Hence, if the *clock-time* intervals in the two co-ordinate systems, between the same two events, are ΔT and $\Delta T'$, respectively, $\Delta T/\Delta T'$ must be a constant quantity; it cannot vary with the particular events chosen for the comparison.

From this it follows inevitably that the co-ordinate differences, Δt and $\Delta t'$, to which the Lorentz transformation applies, cannot, as Einstein's theory requires, represent the clock-time intervals, in the respective co-ordinate systems, between the same events. For, according to the Lorentz transformation,

$$\frac{\Delta t'}{\Delta t} = \frac{1}{\alpha} - \frac{v}{c^2 \alpha} \frac{\Delta x}{\Delta t},$$

where $\alpha = \sqrt{1 - v^2/c^2}$. We therefore get different values for this ratio according to the events we choose, Δx being the space interval and Δt the co-ordinate time-interval, in one of the systems, between those events. We can, by a suitable choice of events, make $\Delta t'$ greater or less than Δt, by any amount we like. But, as we have seen, the relative rates of the *clocks* must be independent of the events chosen for their determination. Hence $\Delta t'/\Delta t$ cannot represent the relative rates of the clocks.*

To see the point quite clearly, suppose that at a certain time I fire a gun. Two persons, R and L, at rest with respect to me, stand at equal distances on my right and left, respectively, and after, say, five minutes by my clock (and theirs) they both simultaneously fire guns. Then my time-interval between my shot and R's is the same as that between my shot and L's—not merely *equal* to it but actually the *same* duration. What is this duration by a moving clock? According to Einstein's formula, the interval between my shot and R's is less than five minutes, and that between my shot and L's is more than five minutes. Hence the moving clock goes slower or faster than mine according to the positions of the events by which I make the test. (Δt is the same

APPENDIX II 273

in both cases, but Δx has a plus sign in one case and a minus sign in the other. The other quantities in the equation are constant, and Δx can be chosen so that $\Delta t'/\Delta t$ is greater and less than unity in the two cases). But there is nothing whatever to tell me which events I ought to choose. Nature knows no distinction between left and right, and Einstein's theory implies that there is no distinction.

It is therefore impossible that the co-ordinate time intervals, Δt and $\Delta t'$, can indicate the time intervals recorded by clocks as the theory requires.

---

* From this point onwards the text differs from that submitted to *Nature* (see p. 73), since the latter contained technical references which would be out of place here.

INDEX

Aberrational constant, 132
Absolute, the, 183
Acceleration, 71
Achilles, 77
Acoustics, 60, 66, 113, 186, 199, 210
Action, and experience, 251-4; and ideas, 14; and reason, 200-1
Action at a distance, 58, 59, 88-9, 212, 216, 232, 265-6
Adrian, Lord, 180, 224
Advisory Council on Scientific Policy, 224
Alchemy, 186, 208
Amsterdam Academy, 80
Analysis, 83
Andrade, E. N. da C., 116
Animism, 24
Anthropocentricism, 43
Antisthenes, 214
Aristotle, 15, 128, 180, 189
Arithmetic, 145-6
Art(s), 46, 196
Asquith, 7, 10
Astronomy, 18, 43, 51, 52, 63, 120-4, 126, 130, 134, 215, 216, 230
Astrophysical Journal, 69
Asymmetrical ageing, 65, 70-2, 76, 77, 79-82, 85, 89, 143, 151, 152, 158, 233
As You Like It (Shakespeare), 231
Atom, stability of, 125; *see also* Particles
Atomic energy establishments, 174, 175
Atomic research, 61, 74-5, 153, 174, 232
Atomic theory, 27, 29-30, 51, 153, 175-6, 209, 210
Augustine, St, 255

Bach, 244
Bacon, Francis, 96, 105, 214
Balliol College, 222
B.B.C., 79, 268-9
Becquerel, 51
Beethoven, 244
Belief and Action (Samuel), 8
Bell, E. T., 137
Bergson, 179
Berkeley, 176n
Beveridge, Lord, 268
Bible, 226
Biem, W., 80

Biochemistry, 178
Biology, 29, 139, 178, 186, 187, 204, 208, 222, 224, 236
Biophysics, 178
Birmingham, University of, 227-8
Body, functioning of, 178; and mind, 179-80, 187-9, 236-7, 242
Bondi, Prof. H., 79, 120, 143n, 239, 240
Born, Prof. Max, 80, 153, 154, 156, 157, 170, 173, 232, 240
Boston, 222
Boswell, James, 176n, 218
Brain, Sir Russell, 27-8
Britain, and ICSU, 246, 247, 248n
British Academy, 248n
British Association for the Advancement of Science, 225, 234, 235
British Journal of Philosophy and Science, 261n
British Society for the History, Philosophy, of Science, 247, 248
Broadcasting, 51, 68-9
Broglie, Prof. Louis de, 135
Browning, 238, 239
Buddhism, 255
Budget of Paradoxes (de Morgan), 138
Buridan's ass, 77, 81

'Caloric', 117, 192
Canterbury, Archbishop of, 228, 257
Cardiff, 234
Carlyle, 241
Carroll, Lewis, 218-19
Catherine the Great, 137-8
Causality, causation, 114-15; and chance, 153, 154, 155, 156, 157, 232; and phenomenon, 100, 103
Cavendish, 166, 169
Cell, study of, 180
Chaldaea, 216
Chance, 153, 154, 156, 157, 232
Chemistry, 29, 224, 230
Cherwell, Lord, 54, 57, 267
Chestnut tree, 206-7
China, 25, 43, 216
Christianity, 225, 226, 228, 244, 254, 255, 256
Church of England, 228
Clapeyron's equation, 112
Clarkson, 263

INDEX

Cleopatra, 109
Colour, 209
Commonwealth, British, 223
Communism, 181, 204
Concepts, nature of, 235-6
Concise School Physics (Shackel), 116
Concomitance, 236
Conduct, and religion, 255-6; *see also* Morality
Consciousness, 259, 261
Continuous creation, 119-20, 127-30, 132, 134
Copernicus, 211, 263
Cornell University, 265
Cosmology, and ether hypothesis, 234; hazards of, 128
Cosmology, 143n
Creation, continuous, 119-20, 127-30, 132, 134; meaning of, 240-1
Curie, Marie and Pierre, 15, 51, 211
Cybernetics, 181
Cytology, 180

Dams, 93-5, 98, 102-5
Darwin, 15, 211, 225
Definitions, 182
Deity, 183, 184, 195, 196, 204, 205
De Morgan, 137-8
Descartes, 15
Determinism, 157-8, 252-3
Dialogue on the Two Great Systems of the World (Galileo), 99-100
Diderot, Denis, 137-8
Differential equations, 30, 40-1, 125
Dingle, Prof. Herbert, 9, 15-19, 31-41, 48-50, 61-75, 78-89, 96-101, 108-14, 127-33, 137-49, 158-74, 177, 185-203, 211-13, 237-64, 270-3
Dirac, Prof., P. A. M., 54, 266-7
Discovery, 79n
Don Marquis, 43
Doppler effect, 121, 124, 130-1, 134
Dunne, J. W., 151-2, 159-61

Earth, and gravity, 141, 211, *see also* Gravitation; velocity of, 51, 53, 62, 63, 64, 90
Eastbourne, 179
Ecclesiastes, 10, 26
Eclipses, 25, 42-3, 48, 215, 216
Eddington, Sir Arthur, 27-8, 29, 30, 39, 40, 54, 111, 260, 265
Education, human and technical, 222-3; scientific, 116, 224, 232
Edward Cadbury Trust, 227

Einstein, Albert, 8-9, 47, 52, 54, 58, 62, 65, 66, 67, 68, 69, 70, 75, 132, 153, 158, 216, 232, 233, 266; and Dingle, 72-4, 270-3; and indeterminacy, 154, 156-7, 173
Élan vital, 179
Electricity, and ether, 266; and magnetism, 212; and Maxwell, 62, 63
Electromagnetic field theory, 62, 64, 65, 85, 88, 100
Electromagnetic waves, 42, 51, 55, 58, 232, 267, 268, 269; *see also* Waves
Electromagnetism, science of, 186, 212, 213
Electron, charge of, 169; discovery of, 51, 153; mass of, 167-9; movement of, 63; nature of, 170-2, 174, 176; position of, 169-72; *see also* Particles
Electronic computer, 181, 190-1
Electron-microscope, 180
Elizabeth II, Queen, 223
Ellison, M. A., 122
Emancipation, 203
Emerson, 106, 230
Endeavour, 122
Energy, active and quiescent, 57-8, 59, 87, 89-91, 92, 98, 99, 101, 105-7, 109, 111, 112, 129, 136, 148, 149, 230, 231, 232; concept of, 242, 243; conservation of, 91-2, 96, 97-8, 99, 104, 105, 108, 112, 113, 114, 117, 120, 132, 221, 233; current teaching on, 116; and ether, 55, 57, 85; and hydro-electric dams, 93-5, 100-1; and light, 135; nature of, 85-7, 182, 191-2, 204, 205
Epanimondas the Cretan, 77
Essay in Physics (Samuel), 8, 18, 116, 155n
Ether, Dingle on, 84-9, 101, 111, 112; and Einstein, 65-6, 67, 68, 69; existence of, 37, 51, 53, 54, 57, 58, 60, 63-4, 215, 230, 265-9; multiplicity of, 205, 211-13; Samuel on, 89-91, 96, 106-7, 118, 120, 122-3, 125, 134, 136, 230, 231-4
Ether and Reality (Lodge), 265
Ether of Space, The (Lodge), 212n
Ethics, 14, 200-3, 254, 255-6
Euclid, 145
Euler, 137-8
Events, 271, 272
Evil, problem of, 199
Evolution, 201, 219, 225, 226
Evolution of Physics (Einstein and Infeld), 271

Existence, 240-1; as prior to perception, 23, 34
Existentialism, 254
Expanding universe, theory of, 52, 120-4, 130-3, 134, 233
Experience, and astronomy, 215; and mathematics, 142, 144, 145, 146, 147-9, 150-1, 163-73, 254; and morality, 203; as primary datum, 17-19, 22, 32, 34, 35, 162, 217, 219-20, 238; as only reality, 114, 118, 189-91, 206, 208, 211, 229, 241; and religion, 195-200, 256, 257, 259; and science, 37, 38, 39-40, 41, 43, 48, 49, 129, 138-9, 140, 144, 148-9; and the supernatural, 193-5; and voluntary action, 251-4
Experiment, scientific, 142, 144, 238, 245
Experiment with Time, An (Dunne), 151-2, 159
Eye, evolution of, 126

Faith, 202-3; and science, 185
Faraday, 61, 65, 73, 87, 88, 101, 211
Fechner, 48
Fisher, Dr, 228
Fitzgerald, 63
Force, lines of, 61, 65, 87, 88, 101, 211
Forces, theory of, 97, 101
Frankenstein, 36, 218, 219
Friends, Society of, 257-8
Fréjus dam, 94-5, 98, 102-4, 108, 109, 114, 115
Fresnel, 85
Fuller, Margaret, 199

Galaxies, 120-4, 131-2, 205
Galileo, 96, 99, 103, 137, 211, 263
Gamma-rays, 51
Gases, kinetic theory of, 142
General Science for Schools (Taylor), 116
Genesis, 225, 226, 256, 258
Geology, 35, 219
Gibbon, 26
Gladstone, 226
God, 74, 138, 197, 254-8, 259; dice-playing, 157, 173; in every man, 263-4; nature of, 198-9; and Newton, 88; personal, 184, 257; and science, 226; as will and power, 38-9; will of, 201, 202, 257
Gosse, Edmund, 34-5, 38-9, 219
Gravitation, 32, 39, 58, 141, 164, 165-6, 188, 189, 208, 209-10, 211, 212, 213, 221, 238, 266; and energy, 86, 94, 95, 96, 102-5, 108, 109, 110, 111, 114-15, 233; nature of, 99-101, 103, 104
Greeks, ancient, 25, 43
Gromyko, 181, 189-90

Hamlet, 171
Harvard University, 223
Harwell, 30, 42, 61
Heat, 112-13, 117-18, 126, 186
Hegel, 15
Heisenberg, 152-3, 156, 174, 222; *see also* Indeterminacy
Hertz, 51, 62, 85, 127
Hinshelwood, Sir Cyril, 221-2, 234, 236, 242, 250
History of the Theories of Aether and Electricity (Whittaker), 267
Hitler, 255
Holland, 117
Hoyle, Prof., 119, 250n, 258
Hubble, Dr Edwin, 121, 122
Humanist tradition, 222-3
Humanities, and science, 250n
Huxley, Julian, 116
Huxley, T. H., 74, 225, 226
Hydro-electric, *see* Dams
Hydrogen atom, 167

Ideal concepts, 163-73
Idealism, 20-1, 22, 23, 24, 27, 28, 31, 33, 39, 43, 44, 189
Ideas, confusion of, 8, 13
Illusion, 190
Imagination, 190
Indeterminacy, principle of, 148, 152-3, 156, 157-8, 172-4
India, 25, 225
Individual, 20
Infeld, 271
Inge, Dean, 22, 185
In Search of Reality (Samuel), 9, 18, 206n, 230
Integrative Action of the Nervous System, The (Sherrington), 180
International Council of Scientific Unions, 246
International Union for the History and Philosophy of Science, 246, 248n
Introduction to Mathematics (Whitehead), 147
Israel, 260
Italy, 26

Japan, 52; and atomic fall-out, 30, 42, 174-5, 177
Jerusalem, 7-8, 52

INDEX

Job, Book of, 202
Jodrell Bank, 15
Johnson, Dr, 46, 176, 218
Jones, Sir Harold Spencer, 120
Judaism, 244, 254, 255

Kant, 15, 200
Kapteyn, 271
Keir, Sir David, 222
Kelvin, Lord, 97
Kepler, 211
Kinetic energy, 86, 92, 93, 94, 96, 97, 108, 115, 116, 232, 266
Knowledge, growth of, 21, 23-4, 31-2, 37-8; and language, 182; and science, 235-6

Language, 18, 83-4, 259; see also Linguistics
Lay Sermons (Huxley), 226
Lead, properties of, 209-10, 232
Lear, King, 173
Leukemia, 30, 39-40, 175
Life, and the biologist, 186; mystery of, 179; nature of, 182, 191-2, 204, 205, 206-8, 236; study of, 187-9
Life Force, 179
Life of Samuel Johnson (Boswell), 176n
Light, and Doppler effects, 130-1; and ether, 265, 266; nature of, 37-8, 66, 84, 113, 126, 131, 133, 135, 194; and the red-shift, 120-4; reflection of, 209, 210; speed of, 59, 62, 66, 131, 132; transmission of, 232, 265, 266; see also Action at a distance, Ether, Waves
Lindemann, Prof. F. A., see Cherwell
Linguistics, 15, 18, 71, 83-4, 182, 192, 204, 254
Lion and unicorn, 184n, 196
Loccum, 80
Lodge, Sir Oliver, 54, 55, 212, 265-6
Logic, 83-4, 254
Logical positivism, 71, 182, 192, 204; see also Linguistics
London, schools of, 116, 232; University of, 245, 247
Lords, House of, 224
Lorentz, 63-70, 72, 75, 77, 85
Lorentz transformation equations, 64, 67, 69, 270, 272
Love, 255
Loyala, 201, 203
Luther, 201, 203

McCrea, Prof., W. H., 79, 80, 82n
Mach measurement, 60
Magnetism, 186, 188, 189, 208, 210, 212
Malaria, 25-6
Man, environment of, 220; evolution of, 203; and God, 263-4; and nature, 229, 230-1, 262; and the universe, 22-3
Man-made fibres, 231
Marvell, Andrew, 46
Marxism, 181
Mass, concept of, 140, 141, 163, 164, 166-9; and the ether, 266; and motion, 63
Massachusetts Institute of Technology, 222-3
Materialism, 180-1, 189, 204
Mathematics, 45, 103; and chance, 154; education in, 224; and experience, 142, 143, 146, 147-9, 156, 163-73, 254; as means only, 239-40; metaphysical character of, 150-1; and physics, 63, 64, 137, 139, 140, 144, 145, 149, 158, 162-73, 175-6; place of, role of, 47, 137-8, 153, 221, 222, 236; as substitute for reason, 52, 62, 68-9, 74, 78, 80, 89, 104, 108, 129, 142-3, 145, 146-7
Matter, and energy, 88-90, 135; and mind, 208, 236-7, 242-3; nature of, 29-30, 176, 177, 260-1; 'properties' of, 209-10, 232; study of, 186; theories of, 40
Matthews, Dr W. R., 225, 226
Maxwell, Clerk, 37, 42, 62, 63, 75, 85, 88, 126
Mayhew, Christopher, 181
Measurement, and mathematics, 139, 142, 149, 166, 172
Mechanics, 140, 154
Medicine, 25-6
Mein Weltbild, 266
Memory, 49, 181
Mendel, 211
Meredith, 25
Metal fatigue, 123
Metaphysics, 183, 221
Michelson, 51, 52-4, 62, 63, 90, 128, 153, 205, 231, 232
Milky Way, 121, 205
Mind, and body, 31, 179-80, 187-9, 236-7, 242-3; and computers, 181, 190-1; nature of, 182, 191-2, 204, 205, 206, 207-8; study of, 186, 187, 188
Mind, 17n
Miracles, 193, 194
Mock-turtle, 218, 241

Mohammedanism, 255
Moon, 141, 216
Morality, codes of, 201-3; relaxation of, 185; and religion, 184, 185; and science, 237, 263-4
Morley, 51, 52-4, 62, 63, 90, 128, 153, 205, 231, 232
Moscow, 180-1
Motion, and the ether, 64, 65, 84, 85, 87, 90-1; and mass, 63; relativity of, 233, see also Relativity; supposed properties of, 82
Music, 244
My Philosophical Development (Russell), 22, 156
Mysticism, 197
Myths, religious, 184-5, 196, 226

Natural Philosophy of Cause and Chance (Born), 153, 154
Nature, as experience, 217, 229; laws of, 21, 22, 24, 25, 111, 173, 194; and man, 229, 230-1, 262; and mathematics, 137, 156; and the supernatural, 183, 193; and truth, 25
Nature, 73-4, 79n, 144n, 266
Nature of Experience, The (Brain), 27
Nature of the Universe, The (Hoyle), 258
Nebulae, 120-4, 131-2, 211
Neurology, 179
Neutrino, 97
Newton, 32, 51, 52, 54, 58, 88, 89, 99, 100, 103, 109, 127, 162, 164, 165, 211, 232
Nietzsche, 201
Nuclear research, *see* Atomic research

Objective, and subjective, 19, 22
Observer, The, 250n
Ockham, William of, 39
Ohm, 143
Oppenheimer, Prof., J. R., 120, 134
Optics, 199; *see also* Light
Opticks (Newton), 127
Oxford, 182, 204, 222, 225

Pacific ocean, 30, 174
Palestine, 7, 52
Palomar, Mount, 15, 121, 205
Parliament, 181, 223, 224
Particles, sub-atomic, 30, 39-40, 42, 55, 61, 97; counting of, 170; creation of, 119, 120, 127-8, 129, 132, 133, 136, 232; and ether, 265-6; and indeterminacy, 156; mass of, 167-9; nature of, 167-73, 174-7; and quantum theory, 153; velocity of, 74-5; and waves, 125-7, 133, 135, 136
Pasteur, 15, 26, 211
Patterns, concept of, 90, 107
Paul, St, 255
Phenomenon, and cause, 100, 114
Philosophical Magazine, 65, 72-3
Philosophy, and experience, 217, 219-20; key position of, 14, 262-3; linguistic, 15, 18, 71, 83-4, 182, 192, 204, 254; meaning of, 251, 253, 254; and physics, 71, 72-3, 83-4, 165, 221-2; and religion, 14, 224-8, 235; and science, 14, 71, 72-3, 150-1, 171, 178, 196-7, 220-3, 234-5, 238, 243-50; and the supernatural, 183; today, 13, 14-15; weakness of, 214
Philosophy of Science, 74n
Phlogiston, 117, 192
Photons, 126, 127
Physicalische Blätter, 80n
Physical Society, 73
Physics, abstraction in, 236; chance in, 154; education in, 224; and experience, 148, 165-73; and ether, 213, 230; history of, 186; and mathematics, 63, 64, 74, 137, 139, 140, 144, 145, 146, 149, 156, 158, 162-73, 175-6; and measurement, 87; and mind-matter relationship, 242; and nature of universe, 99, 106; and philosophy, 14, 71, 72-3, 83-4, 165, 221-2; revolution in, 51-2, 62; rigidity of orthodox, 96, 104-5; and space and time, 47, 49; subjective element in, 50
Physics and Metaphysics (de Broglie), 135
Planck, Max, 31, 52, 153, 157, 205, 230
Plato, 15, 244
Poincaré, 64n, 68, 69, 75
Politics, 81
Potential energy, 86, 92, 93, 94, 96, 104, 107, 108, 110, 111, 115, 116, 232-3
Prayer, 197
Precognition, 152, 159-61
Priestley, J. B., 152
Princeton University, 9
Principia (Newton), 88
Principia Mathematica (Russell and Whitehead), 147, 155-6
Principle of Relativity, The, 270n
Process, 209-11
Prospero, 34, 176, 218, 219, 240, 241

INDEX

Psychology, 31, 139, 186, 187, 204, 212, 241-3, 259, 264
Ptolemy, 15

Quakers, 257-8, 263
Quantum theory, 19, 31, 52, 55, 61, 153, 157, 205, 267

Radar, 51
Radiation, and matter, 135
Radio-activity, 30, 51, 174-5
Radium, 51
Rayleigh, Lord, 266
Realism, 20-1, 22, 23, 24, 27, 28, 31, 33, 39, 43, 44, 180-1, 189, 198, 204, 240-1
Reality, experience as sole, 114, 118, 189-91, 206, 208, 211, 229, 241; nature of, 8-9, 189, 190; and scientific concepts, 236
Reason, and action, 200; and experience, 252-3; as inner light, 114; and science, 142, 143, 144, 145, 146-7
Red-shift, 52, 120-4, 130-3, 134, 136, 233
Reith Lectures, 120
Relativity, 47, 52, 53, 61, 62, 65-71, 130, 131, 153, 158, 212, 233, 267; Dingle on, 72-4, 78-82, 270-3; general theory of, 71, 266; principle of, 64n; Samuel on, 75-8; special theory of, 65, 68-9, 131, 270-3
Religion, and experience, 195-200; and language, 18; meaning of, 251, 253, 254, 255, 259; myth in, 184-5; and philosophy, 14, 224-8, 235; and science, 13, 198, 224-8, 235, 243, 251-8, 264; sectarianism in, 7-8, 13; symbolism in, 184
Right and wrong, 201-2
Ritz, 68
Robot, 218
Roman Catholic Church, 228
Romanes Lecture, of 1893, 225
Romans, ancient, 25, 43
Röntgen, 51, 127
Ross, Ronald, 26
Roulette, 154-5
Rousseau, 182
Royal Society, 72, 221, 225, 226, 234, 235, 240, 246-50, 251n, 264
Russell, Bertrand, 22, 27, 28, 145, 147, 155-6, 157-8, 237
Russia, 180-1, 215
Rutherford, 15, 27, 51, 175, 211
Ryle, M., 128n

St Paul's Cathedral, 225
Samuel, Viscount, 7-10, 13-15, 19-31, 41-7, 51-61, 75-8, 89-96, 101-8, 114-27, 134-6, 149-58, 174-6, 178-85, 204-11, 214-37, 267-9
Satan, 202
Satellites, 134, 215-16
Scholasticism, 96, 105, 109, 115, 214, 232
Science, authority of, 81; current position of, 15, 244-50; discovery in, 234; education in, 224, 232; and experience, 37, 38, 39-40, 41, 43, 48, 49, 129, 138, 140, 144, 148-9; history of, 24, 143-4, 186, 194, 208, 244, 245, 246-50, 261; and the humanities, 250n; and language, 18, 96; and materialism, 181; and mathematics, 137-8, 176; meaning of, nature of, 16, 74, 251, 253, 254; method of, 186-9, 238; and morality, 263-4; and philosophy, 14, 71, 72-3, 150-1, 171, 178, 196-7, 220-3, 234-5, 238, 243-50; and religion, 13, 198, 224-8, 235, 243, 251-8, 264; loss of reason in, 19, 74, 82-4; specialization in, 249-50
Science, Religion and Reality (ed. Needham), 261
Science Museum, 247
Scientific theory, nature of, 113-14
Scientist, as superior being, 143-4
Serial Universe, The (Dunne), 152, 159
Shackel, R. G., 116
Shakespeare, 34, 173, 176, 218, 231, 240-2
Shaw, Bernard, 179
Shelley, May, 36, 218
Sherrington, Sir Charles, 179-80, 189, 207, 209, 237
Sirius, 17-18, 43
Snow, Sir Charles, 9n
Society for the Promotion of Christian Knowledge, 228
Solar system, 48, 216, 231
Solipsism and Related Matters (Dingle), 17
Sound, velocity of, 66, 113
Sound barrier, 60
Sound-waves, 210
Space, 28, 46, 47, 49; and ether, 205, 266; a fictional abstraction, 88; measurement in, 67-8; temperature of interstellar, 141, 142
Space-time, 47
Space-travel, 65, 67-8, 70-2, 75-8, 79-82, 89, 142, 143, 151, 152

Spinoza, 15, 244
Sputniks, 134, 216
States, physical, 118
Statistics, 154, 155, 157, 232
Steady state theory, 132-3; see also Continuous creation
Stephen, Leslie, 154
Structure of the Universe, The (Whitrow), 122
Subjective, and objective, 19, 22
Substance, ether as, 91
Sun, 211, 231; eclipses of, 215, 216, see also Eclipses; emission from, 59-60, 123, 134; energy from, 93, 95, 96, 104; and gravity, 141; motion of, 24, 38
Superman, 201
Supernatural, the, 182-3, 193-5, 211
Symbolism, religious, 184, 196; use of, 147

Tables, of Eddington, 27-30, 39-40, 42, 61
Taylor, F. Sherwood, 116
Technology, and humanities, 222-3; and science, 235
Temperature, 140, 141
Tempest, The (Shakespeare), 34, 176
Templeman, Dr Geoffrey, 227-8
Tennyson, 84, 179
Thackeray, 117n
Theology, 13, 185, 227-8, 255, 256, 257
Thermodynamics, 112, 186, 187, 208, 221, 242
Things Around Us (Andrade and Huxley), 116
Thomson, Sir George, 61, 234, 235-6
Thomson, Sir J. J., 15, 27, 51, 54, 55, 61, 106, 175, 211, 232, 266
Through Science to Philosophy (Dingle), 17n, 20, 253
Time, 34, 35-6, 43-6, 47, 48-9; human and cosmic, 231; impossibility of abstract, 151-3, 158-62; and space travel, 65, 67-8, 70-2, 75-8, 78-82; and special theory of relativity, 270-3
Times, The, 43, 81, 95n, 228
Tizard, Sir Henry, 9, 54, 230
Transmutation, of elements, 37; principle of, 127, 135, 136
Transuranium elements, 125
Truth, criterion of, 25-6; 'drift' towards, 38; and experience, 48; and symbol and myth, 184
Turkey, 7, 225

Uncertainty, principle of, 47, 148, 152, 153, 156, 157-8, 172-4
Unified field theory, 189, 212-13
Universe, as it is in itself, 21, 32, 33, 99, 100, 105-6, 111, 113-14, 118, 148, 149, 150, 157, 162-3, 165, 172-3, 189, 191, 195, 203, 206, 217, 220, 228-31, 234, 236, 237, 259-60, 262; continuous creation of, 119-20, 127-30, 132-3, 134; theory of expanding, 52, 120-4, 130-3, 134, 233; as incomplete, 183; man-centred, 206, 228-31, 236, 237, 260, 262; and mathematics, 162, 169, 172-3; pre-human, 22-3, 34, 35, 36, 44, 45, 210-11; real and apparent, 27-8, 33, 36-7, 39, 43-4, 47; size of, 205; ways of describing, 16-19, 20-1, 22, 32-3, 36
Universe in the Light of Modern Physics, The (Planck), 157, 230n
U.S.A., 9, 30, 175, 177, 215, 222-3, 249, 261

Vacuum, 199, 267
Vanity Fair (Thackeray), 117n

War, first world, 7; danger of third, 13-14; and science, 250n
Washington, U.S.A., 268
Water, energy of, 93-6; varying patterns of, 56
Waterston, 143
Watts, Isaac, 262
Waves, and Doppler effect, 130; and ether, 268-9; as recurring pattern, 57, 58; theory of, 30, 42, 47; transmission of, 59-61, 65-6; see also Electro-magnetism, Ether, Light
Wave-lengths, elongation of, 122-3, 224, 134, 136
Wave-particle problem, 125-7, 133, 135, 136, 233
Weltanschauung, 15-16, 17, 18, 19
Where is Science Going? (Planck), 157
Whitehead, Prof. A. N., 147, 155-6, 176
Whitrow, Dr G. J., 122
Whittaker, Sir Edmund, 54, 61, 69, 267
Wilberforce, Bishop, 225, 226
Wilberforce, William, 263
Will, 38-9
Wilson, Mount, 121
Witch doctors, 41
World as I See it, The, 266
World Science Review, 82n

X-rays, 51, 125

THE END

For Product Safety Concerns and Information please contact our EU representative GPSR@taylorandfrancis.com
Taylor & Francis Verlag GmbH, Kaufingerstraße 24, 80331 München, Germany

www.ingramcontent.com/pod-product-compliance
Lightning Source LLC
Chambersburg PA
CBHW060557230426
43670CB00011B/1856